# ISAAC NEWTON

Peter Ackroyd is a prize-winning writer of both fiction and non-fiction. He is the biographer of T. S. Eliot, Ezra Pound, Charles Dickens, William Blake, Thomas More and, most recently, William Shakespeare.

ALSO BY PETER ACKROYD

Brief Lives series

*Chaucer*
*Turner*

Fiction

*The Great Fire of London*
*The Last Testament of Oscar Wilde*
*Hawksmoor*
*Chatterton*
*First Light*
*English Music*
*The House of Doctor Dee*
*Dan Leno and the Limehouse Golem*
*Milton in America*
*The Plato Papers*
*The Clerkenwell Tales*
*The Lambs of London*

Non-Fiction

*London: The Biography*
*Albion: The Origins of the English Imagination*
*The Collection: Journalism, Reviews, Essays, Short*
*Stories, Lectures* (ed. Thomas Wright)

Biography

*Ezra Pound and his World*
*T.S. Eliot*
*Dickens*
*Blake*
*The Life of Thomas More*
*Shakespeare: The Biography*

PETER ACKROYD

# Isaac Newton

## Brief Lives

**VINTAGE BOOKS**

London

Published by Vintage 2007

12

First published in Great Britain in 2006 by
Chatto & Windus

Vintage Books
Random House, 20 Vauxhall Bridge Road,
London SW1V 2SA

www.vintage-books.co.uk

Addresses for companies within The Random House Group Limited
can be found at:
www.randomhouse.co.uk/offices.htm

The Random House Group Limited Reg. No. 954009

A CIP catalogue record for this book
is available from the British Library

ISBN 9780099287384

Penguin Random House is committed to a sustainable future for
our business, our readers and our planet. This book is made from
Forest Stewardship Council® certified paper.

Printed and bound in Great Britain by Clays Ltd, St Ives plc

# Contents

# List of Illustrations

'La Dioptrique' by Rene Descartes (1596–1650). Engraving *c.* 1637. *Bibliotheque de l'Academie de Medecine, Paris, France/Archive Charmet*

The so-called 'crucial experiment' that shows light from the sun being refracted through one prism and then being refracted through another prism. Pen and ink drawing by Sir Isaac Newton (1642–1727). *MS New Coll 361/2 fol 45v. © Reproduced by courtesy of the Warden and Scholars of New College, Oxford*

Title page from Newton's 'Opticks'. *Science & Society Picture Library, Science Museum, London, UK*

Marble statue of Newton by Louis-Francois Roubillac (1702–62). *Trinity College, Cambridge, UK*

Sir Isaac Newton. Artist unknown. *National Portrait Gallery*

Sir Isaac Newton, 1710. By James Thornhill. *Trinity College, Cambridge, UK*

Charles Montagu, First Earl of Halifax (1661–1715). Studio of Michael Dahl the Elder. *Private Collection/ © Philip Mould, Historical Portraits Ltd, London, UK*

The Tower of London and the Mint. 18th-century engraving. From 'A book of the prospects of the remarkable places in and about the city of London' *c.* 1700. *O'Shea Gallery, London, UK*

Iron cast of Newton's death mask. Possibly after John Michael Rysbrack. *National Portrait Gallery*

William Stukeley (1687–1765). Pen and ink on paper, self-portrait, 1735. © *Ashmolean Museum, University of Oxford, UK*

The author and publishers are grateful to the Bridgeman Art Library for help with picture research and for permission to reproduce.

# Chapter One
## A blessed boy

Isaac Newton, the man who more than any other has shaped modern perceptions of the world, was born at 2 a.m. on the morning of Christmas Day, 1642, outside an obscure village in Lincolnshire from a family of undistinguished yeoman stock. He was a premature and sickly infant. Two women from the household were sent to collect certain items for the child but 'they sate down on a stile by the way & said there was no occasion for making haste for they were sure the child would be dead before they could get back'. At a later date Newton informed a relative that 'he had been told that when he was born he was so little they could put him into a quart pot & so weakly that he was forced to have a bolster all round his neck to keep it on his shoulders'.

Yet the omens were good. A birth on Christmas Day, with its obvious connection to the Saviour, was considered to be blessed. Such a birthday was deemed to be the harbinger of great success in the world. Isaac Newton was an only child. His father had died four months before the son was born, and so the omens were good in another sense: a posthumous child was commonly believed to be the recipient of good fortune. In his adult life Newton considered himself to be unique among men, and the circumstances of his arrival in the world must have encouraged that notion. His survival was considered by him to be a miracle, a harbinger of the other miracles he would perform.

He was born in a small house in the manor of Woolsthorpe-by-Colsterworth. An eighteenth-century drawing shows it to be a solid if undistinguished two-storey house of the English type. The drawing also contains a cow, a horse, and a peasant with his cart, depicting in miniature the rural world in which Newton grew up. The house itself was constructed from the grey limestone of the vicinity and included a kitchen, hall, parlour and upstairs bedrooms. The modern visitor will note that the rooms have low ceilings and stone floors, adding to the generally sober if not exactly sombre atmosphere of the dwelling. Newton was born in the bedroom immediately to the left of the staircase.

The house stands on the side of a small valley made by the River Witham, facing west, and looks over the garden that has become celebrated in world history as the site of Newton's falling apple. The tree itself was long ago felled by the wind. The 'manor' consisted of some hundred acres (31.5 hectares) of woods and fields, a patrimony that Newton would one day inherit. William Stukeley, the eighteenth-century topographer, antiquarian and Newtonian enthusiast who collected all the material on his hero that he could find, described the landscape as one of 'very delightful' valleys and 'plentiful' woods. He added that the 'springs and rivulets of the purest water abound' and that the air of the neighbourhood 'is exceedingly good'. This was the area that produced 'the greatest genius of the human race'.

The ancestry of his immediate family gave no inkling of this greatness. Newton could trace his predecessors only as far back as John Newton, who had lived a century before in the village of Westby just a few miles from Woolsthorpe. The Newtons were husbandmen of Lincolnshire stock who by degrees rose painfully slowly in the social hierarchy of the county and had by Newton's time reached the eminence of yeomen. When on occasions we observe the taciturnity and even surliness of Newton, we may recognise the habits of Lincolnshire farming stock.

His own father, also named Isaac Newton, has no great claim to the attention of posterity. He was a yeoman farmer who looked after the estate and had a certain proprietorial care for the tenants who lived in the small cottages within his domain. He left some five hundred pounds in his will, proving that he had maintained the relative affluence of the family. All the evidence suggests, however, that he could not write his name, like the father of William Shakespeare, thus adding to the myth of genius born in the most unpromising circumstances. Newton's uncle and cousin were also both illiterate. It would have been perfectly possible that, in marginally different circumstances, Isaac Newton himself would never have learned how to read or write.

The family of his mother, Hannah Ayscough, was more genteel in its aspirations. It is the usual familial chemistry of male children who go on to distinguish themselves. Hannah's brother was a clergyman who had studied at Cambridge University. He could not have been very successful in his vocation, however, since he was immured in a rectory only two miles away from Woolsthorpe itself. The Ayscoughs, originally from Rutland, were in fact in a state of sliding gentility. The marriage of Isaac and Hannah reflects their meeting with rural stock on its way up. Isaac Newton was the result.

He was baptised on the first day of the new year, 1643, in the family church at nearby Colsterworth. The widowed mother naturally gave the infant her spouse's name of Isaac. The name itself comes from the Hebrew for 'one who laughs', but the young child could have felt little cause for rejoicing. Just three years after his birth, his mother decided to marry again and separate herself from the infant child. She was betrothed to a neighbouring rector, Barnabas Smith, more than thirty years her senior. The Reverend Smith had no very Christian notion of his stepson, however, and as part of the marriage agreement it was stipulated that the young Newton would remain at Woolsthorpe under the care of his maternal grandmother. Hannah would depart to Smith's rectory in North Witham, some mile and a half away. A relative recalled later that Smith gave Newton 'a parcel of Land, being one of the terms insisted on by the widow if she married him'. Smith also agreed to have the manor house at Woolsthorpe repaired and renovated. It seems to have been altogether a businesslike arrangement, and this was not an age in which the susceptibilities of young children were much considered.

So for the next eight years Newton was reared by his grandmother, Margery Ayscough. There is no mention of his paternal grandfather in this context, so it must be assumed that he played no part in the new domestic arrangements. It was an affluent and respectable household, but his mother's absence necessarily had a profound effect upon the young child. It would have been natural for him to have experienced a sense of abandonment, perhaps even of unworthiness. In adult life he manifested great insecurity and an inordinate fear of emotional contact with other human beings; he was also suspicious and secretive, with a great desire for order and security in all aspects of his life. He also had the capacity for great anger and aggression. These are perhaps the marks of one who had once been deeply hurt.

Whatever the truth of his psychology, there can be no doubt that he was a lonely child. He was brought up in a farmhouse some distance from any village, and he would have been prohibited by his grandmother from playing with any of the 'common' children of the cottagers. He was, in other words, thrown back upon his own resources. In later life he was well known for his isolation as well as for his self-sufficiency, characteristics that he acquired at an early stage. It is often said that brilliant mathematicians tend to have solitary childhoods, in which they can explore the visionary world of numbers.

There are only two recollections of this otherwise obscure period in Newton's life. He recalled that his grandmother had claimed the family's affinity with a baronet, perhaps the context for Newton's subsequent aspiration towards gentlemanly status. And in a notebook compiled at a later date he confessed to the sin of 'threatning my father and mother Smith to burne them and the house over them'. The date of this terrible threat is not known but, to gain its full effect, it must have occurred when Barnabas Smith still lived. Such was the ferocious anger of the young boy abandoned and betrayed. It is significant, too, that he did not forget his anger.

And then, quite unexpectedly, his mother returned. Barnabas Smith died after eight years of marriage, and the twice widowed Hannah came back to the family home with three small children. The advent of a half-brother and two half-sisters may not have been greeted by Newton with overwhelming joy. Barnabas Smith left nothing to Newton in his will, but the boy did receive the pastor's library of two hundred books and a large 'commonplace book' in which Newton was to write down many of his early experiments and speculations. He called it his 'waste book'.

He enjoyed his mother's unfamiliar company for only two years, however, since in 1655 at the age of twelve he was despatched to the grammar school in the neighbouring town of Grantham. It was only seven miles away but the distance, in the intellectual history of Isaac Newton, was much greater. He was lodged with a local apothecary, Mr Clarke, whose house and shop were on the high street of the town, beside the George Inn. It was a busier, and noisier, milieu than any he had before encountered. The apothecary was the brother of the usher at the grammar school, so he may have 'taken in' schoolboy lodgers as part of his income. Clarke himself may have been responsible for Newton's early interest in chemical experiment. In his notebook the young Newton began to transcribe recipes and cures acquired from the books that Clarke kept around him. Newton slept in a garret room, probably with one of Clarke's own children, and he carved his name on the walls as well as leaving drawings of birds and ships, circles and triangles.

Even the doodles of Newton have significance. At some point he drew portraits of John Donne and Charles I. His reasons for memorialising the poet are not easy to guess, but it might seem that he had some sympathy with the executed king. It should not be forgotten that the years of his childhood were the years of the Civil War and the Protectorate, but it would be foolish to speculate about the schoolboy's political sympathies. He was a royalist by necessity at the time of the revived monarchy, but there was also a strong spirit of dissent (even Puritanism) in his religious sensibility.

It is possible that he had learned the rudiments of reading and spelling at one of the 'dame' schools in the local villages of Skillington and Stoke Rochford, but the Free Grammar School of Edward VI in Grantham offered him his intro-duction to the world of classical letters. In particular he was taught how to read and write in the Latin language, an indispensable prelude to any scholastic achievement and a necessary accomplishment for any man of learning. Without Latin, the adult Newton would not have been able to reach a European audience. He also acquired a working knowl-edge of Greek, and was systematically instructed in the Scriptures. Since the books he acquired from Barnabas Smith were largely theological in intent, it is possible that his biblical knowledge was already extensive. Common sense would suggest that he devoured whatever volumes were to hand. At school he was trained in handwriting, employing the correct 'secretary' hand, and may even have been given rudimentary lessons in mathematics.

Yet he was not necessarily precocious in the standard curriculum of the school. In his first year he was marked seventy-eighth out of a total of eighty pupils, and a later biographer has remarked that he 'continued very negligent' in his lessons. Like many children of incipient genius, however, he may have considered those lessons to be unnecessary. Those who are busy in one realm of the mind and imagination may ignore all others as unimportant. Yet he did stand out from his childhood contemporaries in another sense. He was endlessly inventive. There are many anecdotes concerning Newton's youth and childhood; some of them are apocryphal, some are hagiographical in intent, and others are plainly incredible. It is important only to note that they accrued around Newton at a relatively early date; he was celebrated by his contemporaries, and in the eighteenth and nineteenth centuries he was considered a magus almost unrivalled. So the stories and legends grew.

An historian of Grantham, recounting the history of the town's most famous inhabitant, recorded young Newton's 'strange inventions, and extraordinary inclination for mechanics'. Instead of playing with the other boys he preferred to construct 'knick-knacks and models of wood in many kinds', for which purpose 'he had got little saws, hatchets, hammers, and a whole shop of tools'. So he spent his time outside his lessons 'knocking and hammering in his lodging room'. It is a clear foretaste of his dexterity and ingenuity as a mechanic and technician in his own laboratory.

As a child he made a wooden clock, and a wooden mill based upon his observations of a new mill being built in Grantham itself. Within the wheel of the mill he placed a mouse, which made it revolve. The clock itself was powered by water, and was considered so reliable that 'the family upon occasion would go to see what was the hour by it'. From the beginning he was preoccupied with time and mechanism. This was nowhere more evident than in his creation of a sundial by calculating the sun's progress and by fixing pegs to the walls and roof of the apothecary's house.

It was so exact that 'any body knew what o'clock it was by Isaac's dial, as they ordinarily called it'. He devised his own almanac, too, with its report of equinoxes and solstices. In much later life it was observed that he looked at shadows in order to tell the time of the day. As a child, too, he had discovered the connection between the measurement of time and the measurement of space.

It is said by the same local historian that his delight in mechanical ingenuity sometimes interfered with his more orthodox studies, and that duller boys were placed above him in the form. Nevertheless his capacity for learning was such that he could 'outstrip them when he pleased'. He did not altogether neglect boyish sports, however, and it is reported that he designed paper kites according to the correct mathematical proportions; he also created candle-lit paper lanterns which, when tied to the tails of the kites, impressed and frightened the local people as 'comets'. He had learned the power of enchantment as well as entertainment.

He also evinced a strongly practical and empirical streak, whereby his calculations and observations were turned into machines for use. William Stukeley observed of these childhood inventions that Newton manifested a 'sagacious discernment of causes and effects' as well as an 'invincible constancy and perseverance in finding out his solutions and demonstrations'. It is perhaps too easy to find the child in the man, but there can be no doubt that from his school years he was a skilled artisan as well as calculator. A note by John Conduitt, a later relative by marriage, informs us that he had 'carpenter's hands'.

On the window ledge of the schoolroom in Grantham he left one memorial of his education. He carved, with a penknife, the words 'I. Newton'. At a later date he also recorded a few random incidents of his schooldays. In his obsessive list of sins compiled at the age of nineteen he recalls 'putting a pin in John Keys hat on Thy day to prick him', 'Stealing cherry cobs from Eduard Storey' and 'denying that I did so' as well as 'peevishness at Master Clarks for a piece of bread and butter'. It is not clear what is most remarkable – his memory of such minor incidents or his belief that he had committed sin as a result of them. We can never overestimate the sensitivity of his conscience or his typically seventeenth-century sense of sinfulness.

On the day of Oliver Cromwell's death a great wind blew through Grantham. The boys were playing at leapfrog and the young Newton 'tho' not otherwise famous for his activity at that sport, yet observing the gusts of the wind, took so proper an advantage of them as suprizingly to outleap the rest of the boys'. It was Newton himself who recalled this episode and observed with some satisfaction that it was one of his earliest experiments. So he was becoming proficient in aerodynamics as well as geometry. He also related an incident when he was kicked in the stomach by a boy marked 'above' him in the class; according to Newton's own memory of the event he promptly beat him, dragged him to the church by the ears and rubbed his nose against the ancient wall. It is an early indication of the violent temper that he generally preferred to keep in check.

Among the household of Mr Clarke, the apothecary, was a young girl who by her own account was the object of Newton's affections. His apparent disinclination towards women in later years is not the best proof of this young tenderness but 'tis said that he entertained a love for her, nor does she deny it'. Miss Storey herself also recalls that he 'was always a sober, silent, thinking lad'. He did not play with the boys 'but would rather chuse to be at home even among the girls, and would frequently make little tables, cupboards and other utensils for her and her playfellows'. She remembers in particular Newton's construction of 'a cart he made with four wheels . . . and by turning a windlas about, he could make it carry him around the house where he pleased'.

His mechanical propensities did not take up all of his time. Naturally enough the young Newton had an earnest inclination towards study. He was in most respects an auto-didact, and he collected books that assuaged his general thirst for knowledge. He owned standard schoolbooks such as Pindar and Ovid's *Metamorphoses*, but there were also popular science textbooks such as John Wilkins's *Mathematical Magic*. In 1659, at the age of sixteen, he signed many of them '. . . meum Isaac Newton'. He had already begun to compile notebooks on his reading. He also used these notebooks to jot down his other interests, and there are alphabetical lists of words under the headings of 'Artes, Trades & Sciences', 'Birdes', 'Cloathes' and others. He already had the desire, or the need, to systematise and organise his scattered knowledge. In the notebooks, too, were jotted down apparently random phrases for translation from English into Latin – 'What else is to dance but to play the fool', 'A little fellow', 'What imployment is he fit for? What is hee good for?', 'I will make an end. I cannot but weepe. I know not what to do.' You do not have to be a child psychologist to recognise that these stray phrases spring from a somewhat troubled young boy. He may have discerned some power within himself, but he could not then have been sure of its nature. What *was* he to do?

Then, just as his studies began to flourish, his mother called him home. It seems that she believed he had enough book-learning, and must now begin to learn how to run the manor and estate of Woolsthorpe. Clearly it was not the future he had envisaged for himself. There are accounts of his dissatisfaction and restlessness, while the list of the sins he committed in this period tell their own story – 'Refusing to go to the close at my mother's command', 'Striking many', 'Peevishness with my mother', 'Punching my sister', 'Falling out with the servants'. He was clearly a bad-tempered and sometimes even violent child.

He had no interest in pigs, sheep, cattle or arboriculture. There are stories of his aversion, or detachment, which have the stamp of later rationalisation. When his mother asked him to attend to the sheep or to the corn, he obeyed reluctantly and partially. At a later date he told William Stukeley that he preferred 'to sit under a tree, with a book in his hands, or to busy himself with his knife in cutting wood for models'. He was abstracted from rural tasks, and allowed the sheep or cattle to stray. In the manor court he was fined 'for suffering his swine to trespass in the corn fields' and 'for suffering his sheep to break the stubbs' of unenclosed land. His mother asked a servant to watch over him, but the young Newton delegated all his duties to the servant and continued with his reading. He was also very absent-minded. On one occasion he was leading his horse home when it slipped its bridle; he did not notice the animal's absence and walked home with the bridle in his hand. The servants at Woolsthorpe did not think highly of the young master and 'would say the lad is foolish, and will never be fit for business'. He would not have made a good farmer and yet, in slightly different circumstances, that is exactly what he would have been obliged to become.

Yet there were very fortunately those who recognised his ability and his great aptitude for learning. The master of the school, John Stokes, and Newton's uncle, William Ayscough, attempted to persuade Hannah Smith 'what a loss it was to the world, as well as a vain attempt to bury so extraordinary a talent in rustic business'. There is also a story, no doubt apocryphal, of a stranger who, questioning the young Newton on mathematical matters and discerning his latent genius, persuaded his mother to continue his education. Such semi-miraculous interventions are common enough in the hagiographical accounts of stirring men of power.

Stokes himself promised Hannah that on Newton's return to school 'he would make a compliment to her of the 40s per annum paid to the schoolmaster by all foreign lads', a material bounty to expedite the clever boy's progress. She was sufficiently convinced to send him back to the school in Grantham, where he might prepare for the rigours of Cambridge University. He boarded with Stokes himself, and completed the school learning that he would require. From this period may date his purchase of a Greek lexicon and a book of annotations on the New Testament. These were an indispensable preparation for the more gruelling studies ahead.

His re-entry into school life seems to have kindled his wits as well as his scholarly ambition, and he told John Conduitt, the husband of his niece, that 'his genius now began to

mount upwards apace & to shine out with more strength &
as he told me himself, he excelled particularly in making
verses'. These were likely to have been classical verses in
translation, since he never thought highly of English poetry
itself. In fact he 'exceeded the most sanguine expectations
his master had conceived of him'. Such was his prowess that,
on leaving day, with tears in his eyes, Stokes delivered a
speech in praise of his favourite pupil and urged the other
boys to follow his good example.

## Chapter Two
# College boy

In the summer of 1661 he was admitted as a sub-sizar, and then as sizar, to Cambridge University. The chosen college was Trinity, but whether this was his own choice is another matter. The guiding hand behind his placement there seems to have been Humphrey Babington, the brother of the Grantham apothecary's wife and the uncle of Newton's apparently first and only love, Miss Storey. Babington was the rector of the nearby hamlet of Boothby Pagnell but, more importantly, he was a senior Fellow of the college. He also seems to have been convinced of Newton's intellectual potential, and therefore went to some trouble to have the boy admitted to his own college. It seems likely in fact that Newton became 'sizar' or servant to Babington himself, on the rector's infrequent visits to the college, and thus avoided the usual menial duties of that lowly position in the college hierarchy. The word itself perhaps derives from the 'size' of the portions of bread and drink he received as a recompense for running errands and waiting at table.

His name was placed in the matriculation book of Cambridge University on 8 July 1661, as well as those of sixteen others who had also been accepted at Trinity College. He had arrived at the institution where he would remain for the next thirty-five years and where he would write the book that changed our perception of the world.

He brought with him a chamber pot, as well as a lock for his desk, a pound of candles for his studies at night and a quart bottle of ink. He also purchased a 'table [notebook] to jot down the number of my cloathes in the wash'. Newton was always meticulous in his calculations.

Trinity College itself was then the largest, and perhaps the most picturesque, of all the Cambridge colleges. It was described in the seventeenth century as 'the Stateliest and most uniform Colledge in Christendom'. It had some four hundred members, and claimed three of the five Regius professorships at the university. Cambridge itself was not then renowned as a haven for scholarship of any kind. The dons were generally characterised as drunken or misanthropic, while the students themselves were content to pick up only the rudiments of learning. One undergraduate of the period noted that 'I had none to direct me, what books to read, or what to seek, or in which method to proceed'. Newton rescued himself from that dilemma with his own autodidactic genius. The beneficial aspect of this decay of learning lay in the fact that a clever or eager student could move forward in whatever direction he chose.

The university was situated in what was little more than a dirty and ill-kept village. It had not yet shaken off its medieval inheritance, when it had emerged as part of the educational programme of the monastic and clerical orders. The first college, Peterhouse, had been founded in 1284 by Hugo de Balsham, bishop of Ely, and the ecclesiastical connections with Cambridge continued. In 1318 Cambridge received formal recognition as a *studium generale* from Pope John XXII. It was an offspring of the Church. The allegiances had changed at the time of the Reformation, of course, and by the seventeenth century there was a marked Puritan element in the university, but it was still a primarily religious institution. The Fellows were invariably obliged to take holy orders, a stipulation that did not change until 1871.

The university curriculum itself was still enmired in the Middle Ages, too, with the preponderance of lessons devoted to the works of Aristotle. Newton was obliged to study Aristotelian logic, Aristotelian ethics and an antiquated rhetoric that curtailed rather than encouraged inventive expression. In Aristotle's philosophy, too, the earth remained at the centre of the cosmos.

Newton's own early studies are well enough known. He purchased a notebook and transcribed some sentences from Aristotle in Greek. He read the standard textbooks, but from his notes it is clear that he never finished them. He was bored by them. At some point he wrote down a salient phrase in Latin which, translated into English, declares that 'Good friends are Aristotle and Plato, but a greater friend is truth'. It was that 'truth' which he now assiduously pursued.

He was a little older than his contemporaries, as a result of his interrupted education at Grantham, but he would in any case have remained a solitary figure among the generally careless and feckless students of the university. He is known to have formed a friendship with only one of his fellows, John Wickins, a student at Trinity College who was not on good terms with the young man with whom he shared a college chamber. According to his son, Wickins 'retired one day into the Walks, where he found Mr Newton solitary and dejected; Upon entering into discourse they found their cause of Retirement the same, & thereupon agreed to shake off their present disorderly Companions & Chum together'. This is a record of two lonely young men, of retiring disposition, each desperate for a companion. Wickins and Newton retained each other's friendship for twenty years, and for much of that time Wickins acted as his assistant and amanuensis. Nothing is known of their relationship, however, except for the account of their meeting and a short letter that Newton wrote to him in later life. The letter concerned the distribution of Bibles, and concluded: 'I am glad to heare of your good health, & wish it may long continue. I remain . . .' Wickins's son eventually put down some anecdotes of Newton's life in college, but any evidence of affection between the two young men has disappeared.

In the first full year of residence Newton drew up in shorthand the list of sins that has already been noticed. Among the offences no doubt committed after his arrival at Cambridge were 'not living according to my belief' and 'setting my heart on money learning pleasure more than Thee'. The contrition over 'money' may be related to the fact that the young Newton acted as a part-time usurer or moneylender in the college itself. He lent money to fellow sizars and to 'pensioners', who had a higher status in the college. He kept careful note of the sums involved, and marked repayment with an 'X'. Money-lending was not an unusual practice, in all classes of society, but there was still a certain stigma attached to the activity. Among rules drawn up to be observed by scholars was one stating, 'doe you neither lend, nor borrow any things of any Scholler'. Newton was not likely to make himself popular. It is improbable, in any event, that he courted popularity. His niece in later years recalled him stating that 'when he was young & first at university he played at drafts & if any gave him first move was sure to beat them'.

He had several notebooks for his several studies, and it soon becomes clear that Newton was eager to strike out on his own. He left some blank pages and then began a section that was entitled 'Quaestiones quaedam Philosophicae'. For these philosophical questions he used a new handwriting, 'roman' rather than 'secretary', which improved upon the elegance of the hand he had learnt in Grantham. It is also a mark of his new identity as a scholar, as a young man dedicated to the pursuit of learning. He set out several headings, from 'Air' and 'Meteors' to 'Motion' and 'Vacuum' and 'Reflection'. There are some seventy-two topics in total for his investigation, clearly laid down with the purpose of mastering the whole field of natural philosophy. Nothing less would satisfy him. In these early lists he manifests both his preoccupation with systematic knowledge and the desire to fashion all aspects of learning within a unified field. Most of his notes are derived from the books he was then reading, but he also summarises possible experiments – 'To try whether the weight of a body may be altered by heate or cold'.

His eager and alert mind shines through in these 'Quaestiones' where he jots down his notes from Galileo and Robert Boyle, Thomas Hobbes and Joseph Glanvill, Kenelm Digby and Henry More. Henry More himself had the advantage of also being born in Grantham, and had taught the brother of the apothecary with whom Newton lodged as a schoolboy. The two illustrious children of Lincolnshire did eventually meet and it is clear that Newton read More's works with interest. More was one of the 'Cambridge Platonists', a small group of scholars and poets who tried to unify modern scientific learning with a broadly neo-Platonic philosophy of the soul.

Newton's own account of these early university years, given in a letter written more than thirty years later, recalled that he had read 'Schooteen's Miscellanies & Cartes's Geometry' as well as 'Wallis's works'. He was in other words consumed by arithmetic and geometry. Descartes's *La Géométrie* is perhaps still well known but the other two 'works' can be explicated as Wallis's *Arithmetica Infinitorum* and a compendium of Schooten's *Exercitationum mathematicarum* and *Geometria*. Newton was placing himself at the forefront of mathematical research.

One of his later disciples, Abraham de Moivre, asked him about these years, and was given a more anecdotal answer. In 1663 Newton

> being at Sturbridge fair bought a book of Astrology to see what there was in it. Read it till he came to a figure of the heavens which he could not understand for want of being acquainted with Trigonometry. . . . Got Euclid to fit himself for understanding the ground of Trigonometry. Read only the titles of the propositions, which he found so easy to understand that he wondered any body would amuse themselves to write any demon-strations of them.

He told the same disciple that he read Descartes's *Geometry* by degrees, reading ten pages at a time and then going back over them again to see if he had mastered them.

The description is a fair indication of his temperament. He progressed methodically, moving from the less to the more difficult, but he was also capable of those sudden moments of insight or comprehension that rendered Euclid 'easy' to him at first acquaintance. One disciple wrote at a later date that Newton 'could sometimes see almost by Intuition, even without Demonstration'. He was also teaching himself, as any genius must do, rather than relying upon the precepts of his masters. In his first year of prolonged study he assimilated all of the mathematical learning available to him, and was ready to move ahead. Certainly he had enough self-confidence in his own power to write in the margins of Descartes 'Error – error non est Geom'. He had reached the peak, from which he could glimpse the unknown lands. He would be deeply involved in mathematical research, with occasional interludes, for the next thirty years.

It might seem that the senior mathematician at the college, Isaac Barrow, had been given a thankless task in endeavouring to tutor Newton. They first met when Newton was being examined for a scholarship, and was questioned about Euclid by Barrow. Newton's answers were not considered satisfactory, but he was awarded the scholarship anyway – perhaps as a result of some discreet pressure from his mentor, Humphrey Babington. He received a stipend from the college, and the way was prepared for his degree and subsequent fellowship. It seems that Barrow omitted to question Newton about his study of Descartes and, if he had, he might have had some inkling of how far this young man had advanced.

Barrow was the first Lucasian Professor of Mathematics at the university, and Newton duly attended his lectures on that subject. It was perhaps inevitable that the incipient genius 'found he knew more of it than his tutour', according to John Conduitt, 'who finding him so forward told him he was going to read Kepler's Opticks to some gentlemen commoners & that he might come to those lectures'. Characteristically enough Newton immediately absorbed the book, and seems not to have attended the lectures.

Barrow became one of Newton's early champions, however, and was perhaps the first person in authority to recognise his genius. Stukeley records Barrow's confession that 'still he reckon'd himself but a child in comparison of his pupil Newton. He faild not, upon all occasions, to give a just *encomium* on him, and whenever a difficult problem was brought to him to solve, he refer'd them immediately to Newton.'

Newton was writing his first mathematical essays in the summer of 1664, a harbinger of the profound and astonishing work that he would complete in the next two years. Then in the winter of 1664 he wrote down a series of 'Problems', twelve in all, which he set about resolving in the course of the next year.

But it is an indication of the range, as well as ambition, of Newton's genius that he was at the time making preliminary experiments on the nature of light. They had perhaps been prompted by his reading of Kepler's *Opticks*, recommended by Barrow, but almost at once he followed his own path of research. He read Robert Boyle's *Experiments & Considerations touching Colours*, just newly published, and made many notes upon it. He stared at the sun with one eye, to discover the consequences. He was reckless of his own sight in the process, and had to spend three days in a darkened room in order to recuperate from the experience. At a later date he wished to test Descartes's theory that light was a 'pressure' pulsating through the ether. He inserted a bodkin or large needle within his eye 'betwixt my eye and the bone as near to the backside of my eye as I could'. He did this in order to alter the curve of his retina and observe the results. His passion for experiment was such that he risked blinding himself in order to pursue his researches. He was single-minded to the point of obsession.

In this year he also purchased a prism from Sturbridge Fair in order to continue with the fervent study of light that became, with his mathematics, the principal object of his work. The fair itself was held just outside Cambridge, and was a centre for the sale of toys, curios, books and wonderful objects of all kinds. With the prism he was intent upon testing

> the celebrated *Phaenomena of Colours*. And in order thereto having darkened my chamber, and made a small hole in my window-shuts, to let in a convenient quantity of the Suns light, I placed my Prisme at its entrance, that it might be thereby refracted to the opposite wall. It was at first a very pleasing divertisement, to view the vivid and intense colours produced thereby.

His observations, and his meditations on his observations, would change the understanding of light itself. In that same year, too, he became fascinated by cosmology and at a later date told Conduitt that 'he sate up so long in the year 1664 to observe a comet that appeared then that he found himself much disordered & learned from thence to go to bed betimes'. He did not in fact keep to that valuable lesson, and spent many nights at his labours. But the spectacle of this young man, surveying and mastering all at once the realms of mathematics, optics and cosmology, is little short of astonishing.

It is worth noting that, in the examinations for his BA, his results were not equally astonishing. According to William Stukeley, 'when Sir Is stood for his Bachelor of Arts degree, he was put in second posing, or lost his groats, as they call it, which is looked upon as disgraceful'. The 'groats' were small coins left for the examiner by the examinee; if the student does relatively badly, as Newton seems to have done, the coins are forfeited. The explanation is not hard to find. Newton hardly bothered with the ordinary curriculum, and at the last minute 'crammed' the orthodox textbooks so that he could at least pass each test in turn. His mind, and imagination, were elsewhere.

## Chapter Three
# The apple falls

It is perhaps not unexpected that the years from 1664 to 1666 would be described in retrospect by historians of science as the '*anni mirabiles*' of Newton's life. In these years, according to his own account, he made many advances in mathematics beyond the reach of any of his contemporaries. He discovered what he called 'the method of flowing quantities, or fluents' better known now as integral calculus, and he hit upon 'the calculus of fluxions' or differential calculus. He also 'had the Theory of Colours' and 'began to think of gravity extending to the orb of the Moon'. In addition: 'I deduced that the forces which keep the Planets in their Orbs must be reciprocally as the squares of their distances from the centers about which they revolve.' He was, in other words, on the threshold of the great revolution in human thought that has become known as the Newtonian Revolution. The mysteries of light and gravitation were slowly being revealed to him. As he added parenthetically: 'in those days I was in the prime of my age for invention & minded Mathematics and Philosophy more then [than] at any time since'.

His meticulous financial accounts at Cambridge reveal that, in this prime of life, he was by no means extravagant. There are records of payments for gloves, stockings and a hatband as well as miscellaneous expenses for 'cherries,

Tarte, Custourde, Herbes & washes, Beere, cake, Milke'. There is also reference to 'teniscourte, Wine little, chessmen', which suggests that he engaged in a little sporting competition with his friend Wickins.

There was an unfortunate interruption in this university existence, however. In June 1665 he was obliged to leave Cambridge because of the onset of the Great Plague. In this month the pestilence, or 'the death' as it was commonly known, had reached London from the western suburbs; the grass was beginning to grow in the abandoned streets of the city, and Cambridge was within easy reach. The annual fair at Stourbridge was cancelled; the colleges were closed down; the poor, living in overcrowded tenements, were the inhabitants who most grievously suffered. So Newton retreated to the family home of Woolsthorpe, from which safe distance he continued with his studies. He had brought his books with him, and even put up new shelves to accommodate them.

He also made use of the library in the parsonage of Humphrey Babington in the adjacent parish of Boothby Pagnell. In a later memorandum Newton confirmed that 'I computed the area of the Hyperbola at Boothby in Lincolnshire to two and fifty figures'. It seems likely that Babington had some inkling of the mathematical genius of the young man under his roof. The labour was immense, the result startling. Newton declared that 'I keep the subject constantly in mind before me and wait 'til the first dawnings open slowly, by little and little, into a full and clear light'. His intellectual energy was matched only by his assiduity. There also seems to have been a pattern of progress in his studies. He devoted all of his attention to a problem until he had satisfactorily resolved it; then he would abandon his work for a while. After a few months he would return to it, and make another leap forward. He learned how to 'husband' his mind, as it were, to allow it to lie fallow before it became fruitful once more.

He returned to Cambridge in March 1666, in the false belief that the plague was abating, and then left once more for home four months later. Here he remained for a further ten months, and in this period declared himself to be 'Isaack Newton of Wolstropp, Gentleman, age 23'. The title of 'Gentleman' was no mere honorific; Newton was claiming 'gentle' status as lord of the manor and bachelor of arts at Cambridge University. The manor house of Woolsthorpe seems to have been in fact the setting for Newton's most famous observation during this period of the plague. The story of the apple has entered English lore, despite or perhaps because of the fact that it remains unproven.

There are four separate versions of the apple falling from the tree, in the orchard of the manor house, for the simple reason that Newton recounted different versions to four separate people. William Stukeley lent the story a personal tone. 'After dinner', he wrote of an occasion towards the end of Newton's life,

> the weather being warm, we went into the garden and drank tea, under the shade of some appletrees, only he and myself. Amidst other discourse, he told me, he was just in the same situation, as when formerly the notion of gravitation came into his mind. It was occasion'd by the fall of an apple, as he sat in a contemplative mood.

The account given by Newton's relative, John Conduitt, reveals that

> while he was musing in a garden it came into his thought that the power of gravity (which brought an apple from the tree to the ground) was not limited to a certain distance from the earth but that this power must extend much farther than was usually thought. Why not as high as the moon said he to himself & if so that must influence her motion . . .

It has been understood, ever since the circulation of these stories, that the theory of universal gravitation occurred to Newton as he sat musing in the garden. The legend is implicitly related to the tree of knowledge, and the eating of the forbidden fruit, in the Garden of Eden. The image of the garden is in any case dear to the English. The intervention of nature in the workings of genius is considered to be a particularly beneficent sign.

Newton himself seems to have encouraged this pious misunderstanding of the falling apple. But it did not happen like that. The evidence of his papers confirms that there was much labour of thought and of calculation ahead of him. There were, for example, problems of circular motion and centrifugal force to be solved. Yet some such incident did remain in his memory for many years, and it is possible that the fall of the apple prompted speculations in him that could not immediately be resolved.

It is not coincidental that in this period he returned to mathematics, and compiled a large document with the title

'To resolve Problems by Motion these following Propositions are sufficient'. It has the merit of being the first written work on the calculus, but it is one that Newton kept to himself. After completing the tract he effectively left mathematics alone for the next two years. The fact that he was now the leading mathematician in England, and perhaps in Europe, was known only to himself. There may have been another sentiment at work in his most complex nature. To know something that no one else in the world knew or understood – that was a most exhilarating experience of power. Perhaps he wished to prolong it for as long as possible.

There was another reason for his withdrawal from mathematics. The twenty-three-year-old was now close to formulating a theory of colours that would revolutionise the discipline of optics. He studied the nature of reflection, and of the refraction of light from a variety of curved surfaces. He believed that light itself comprised 'corpuscles' in constant motion, but his preoccupation was with the nature of colour itself.

It was here that he made his breakthrough. By carefully analysing and investigating the effects of the prism, he concluded that white light itself was not some basic or primal hue but was instead a mixture of all other colours in the spectrum. White light was heterogeneous, in other words, and the other colours emerged from it in the process of refraction. The individual rays of light, integral and immutable, excite sensations of colour when they strike the retina of the eye. His conclusions were contrary to tradition, and indeed contrary to the precepts of common sense, but he had absolute faith in his experiments. He also applied himself to mechanical tasks, and in this period 'I applied my selfe to the grinding of Optick glasses of other figures than *Spherical*'. This would later result in his construction of one of the first reflecting telescopes.

One of his disciples stated later that, during this work on optics, 'to quicken his faculties and fix his attention, he confined himself to a small quantity of bread, during all the time, with a little sack and water, of which, without any regulation, he took as he found a craving or failure of spirits'. It is an indication of his absolute and single-minded determination to complete the task he had set himself, yet of course it also promotes the image of the abstemious saint or hermit.

Newton's rate of progress and of discovery in these matters cannot satisfactorily be resolved. Exact measurements of his rooms at Trinity, and of his study at Woolsthorpe, suggest that his experiments with the prism could have been undertaken at either location. The palm may be awarded both to Lincolnshire and to Cambridgeshire, perhaps, as the birthplace of modern optics. All that can be claimed with certainty is that he was ready to lecture upon what he called 'the celebrated Phaenomena of Colours' by the beginning of 1670. This much can be said of his enforced withdrawal from Cambridge. Between 1665 and 1666 this young man revolutionised the world of natural philosophy. He gave the first proper treatment of the calculus; he split white light into its constituent colours; he began his exploration of universal gravity. And he was only twenty-four years of age.

In the spring of 1667 he returned to Cambridge where he would complete his progress towards the degree of Master of Arts. He had been spared the plague, but his tendency to hypochondria was not noticeably altered by this happy escape. He was also climbing the hierarchy of the college. In the autumn he was elected a minor Fellow, obliged to swear 'that I will embrace the true religion of Christ with all my soul'. He would often consult his conscience, and his books, about what that 'true religion' actually was. He was given a stipend of two pounds a year, and a new lodging next to the chapel.

He returned to Woolsthorpe at the end of the year, and received from his mother thirty pounds with which to buy the new clothes and furniture appropriate to his status. He was in fact assured of a relatively comfortable income. He collected rents from the property he had inherited in Lincolnshire, and he was also granted an allowance from his mother. On his return to Cambridge, after two months at home, he purchased a suit and consulted a shoe-maker. Previously he had paid two pounds for two yards (1.8 metres) of cloth '& buckles for a Vest'. Altogether, over two years, he spent the large sum of twenty pounds on his clothing. He was, in other words, careful of his appearance. This appears to be at odds with the image of the absent-minded genius, or indeed of the cloistered hermit. But the fact that he allowed himself to be painted on many occasions, in subsequent years, suggests that he was proud of his image as a scholar and as a gentleman.

He also hired a joiner and a painter to refurbish his chambers while at the same time purchasing a leather carpet and a couch. With the first onset of success had come a natural desire to 'show off' his prosperity. He also seems to have made some effort to share the society of his colleagues.

In his account books he records sums spent in various taverns, as well as fifteen shillings lost in playing cards. This is not at all the conventional image of Newton, rapt and remote, sailing, in Wordsworth's words, through 'strange seas of Thought'; yet he always understood the use and value of money, and in this period he continued his practice as a usurer or moneylender.

In the summer of 1667 the Dutch fleet invaded the Thames and sailed so far up the river that their guns could be heard in Cambridge. Newton then informed his colleagues that the Dutch had beaten the English and, when he was asked how he could form such a judgement, he replied that 'by carefully attending to the sound, he found it grew louder and louder, consequently came nearer; from whence he rightly infer'd that the Dutch were victors'. It is a signal instance of the acuity of his senses, and of his ability to deduce general laws from observation.

Soon after his return from Woolsthorpe he was awarded the degree of Master of Arts and elected as a major Fellow of the college, in which capacity he was paid an annual dividend from the college endowments. Then, in the summer of 1668, he made his first visit to London. He spent a month in the capital, although the nature of his activities is quite unknown. He could not have spent much time in taking in the 'sights', since much of the city had burned down two years before. By the time of his journey the contours of the old streets had been marked out for rebuilding, and twelve hundred houses had already been erected, but Newton was essentially visiting an enormous building site. For one of his inquisitive and practical disposition, however, that may have been of sufficient interest.

It is also likely that he visited instrument-makers and glass-grinders, since on his return to Cambridge he began to construct a telescope. When at a later date John Conduitt asked him where he had had it made, he replied that he had made it himself. Then his relative asked him where he had obtained his tools for this difficult enterprise, and Newton replied that 'he made them himself & laughing added if I had staid for other people to make my tools & things for me, I had never made anything of it'. Here is an experimenter who proceeded on the basis of complete self-reliance. From his accounts it is evident that he purchased a 'Lathe & Table, Drills, Gravers' and assorted other tools. He made a parabolic mirror from an alloy of tin and copper that he himself had devised. He smoothed and polished it so that it shone like glass. He built the tube, and the mounting.

From his experiments in optics he knew that a reflecting telescope would be more effective than the conventional refracting telescope, since the parabolic mirror would obviate the distortions of light resulting from the use of lenses, and in fact his six-inch device had a power equivalent to a six-foot refractor. If he had only ever achieved this one feat alone, he would be worthy of the highest praise. He wrote to a friend at the beginning of the following year in a mood of triumph, explaining that his new telescope magnified 'about 40 times in diameter'. He added that 'I have seen with it Jupiter distinctly round and his Satellites, and Venus horned'.

## Chapter Four
# The darker art

The invocation of horned Venus leads inevitably to the great study that Newton began to pursue in this period. It may have been one of the reasons for his visit to the capital, and for his long stay there. It will certainly account for his purchase of small furnaces, alembics and other vessels. The young Newton had become enamoured of alchemy.

In the vulgar mind the alchemist was only intent upon creating gold out of lesser metals, a miracle of transubstantiation that would make its operators rich beyond measure. That is why the monarchs of Europe welcomed various alchemists to their courts. Yet for Newton, and for other adepts, the goal was spiritual rather than material. 'Alchemy tradeth not with metals as ignorant vulgars think', he wrote. 'This Philosophy is not of that kind which tendeth to vanity & deceipt but rather to profit and edification inducing first the knowledge of God.' In re-creating the substance of the world and producing gold, the alchemist was re-creating himself in the image of the godhead. In one of his notebooks he jotted down a few words: 'Sowing Gold into the earth; Death & Resurrection'. The womb of the earth might bring forth new life.

Like other alchemists he believed that the universe was instinct with life and spirit; it was not simply the collection of lifeless corpuscles or atoms favoured by mechanical philosophers. He believed in the feminine and masculine 'semen' that was responsible for generation and new life. He stated in one of his short papers that 'the vital agent diffused through everything in the earth is one and the same. And it is a mercurial spirit, extremely subtle and supremely volatile, which is dispersed through every place.' This is not so far distant from his theory of universal gravity, yet to be formulated, and it has often been noted that his concept of 'occult' forces in the material world – such as the evident 'attraction' and 'repulsion' between particles that no one could explain – helped to create the arguments of the *Principia Mathematica*.

All the ardour of Newton's nature went into his new pursuit. One of the books he purchased in London was Lazarus Zetzner's *Theatrum chemicum*, an anthology in six volumes of treatises on the recondite art. He also bought 'Aqua Fortis, sublimate, oyle perle, fine Silver, Antimony' and a variety of other substances. He was intent upon experimenting, and in his chambers at Trinity he fashioned for himself a laboratory.

He had in previous years already begun taking notes on what might now be called orthodox or conventional chemistry, under such headings as 'Amalgam', 'Crucible' and 'Extraction'. But these were preparatory to what he considered to be his great work in the manipulation of the material world. Over the next few months and years he made contact with a largely secret group of alchemical adepts, with whom he exchanged texts and information. There was a bookshop in London, at the sign of the Pelican in Little Britain, that acted as a clearing house for published and unpublished material. Newton borrowed unpublished materials, and carefully noted their contents. He drew an image of Jupiter enthroned with a triple tiara, one of the secret signs of the alchemists. He even devised for himself a pseudonym for his alchemical work – 'Ieova sanctus unus', as a near anagram of 'Isaacus Nevtonus'. The assumption of a name meaning 'the one holy Jehovah' may seem somewhat blasphemous, but it is perhaps indicative of the young Newton's self-belief. Had he not been born, like the Saviour, on Christmas Day?

He approached the study of alchemy with his usual care and thoroughness. He purchased every treatise that he could find, both ancient and modern, and after his death he left something like a million words devoted to the subject. In his library there were some 175 alchemical books, approximately one-tenth of the total. This was not some passing or temporary interest. He did not take it up, exhaust it, and put it down – as he did with optics and mathematics. It was a continuing preoccupation, engaging his attention for over thirty years. He joined other alchemists in the endless and ultimately futile pursuit of the 'philosopher's stone', otherwise known as 'the elixir of life', that might transmute base metals into silver or gold.

From his multifarious notes it is clear that he was attempting to synthesise all previous alchemical experimentation, and to apply to the subject his genius for empirical observation. At a later date he compiled an *Index Chemicus* with almost nine hundred headings in which to schematise his reading. It was not an alternative to conventional chemistry; it represented a deepening, and heightening, of the skills he had learned in the laboratory.

And he did not rest. He was so absorbed in his studies that he often forgot to eat. He did not retire to his bed until the early hours of the morning, and then slept for only five or six hours before springing up to resume his labours. There were periods when he would work in his laboratory for six weeks at a time, never letting the fire fail, so that he seemed to his nonplussed assistant to be aspiring to 'something beyond the Reach of humane Art & Industry'. Indeed he was. When he was an old man, living in London, he told John Conduitt that 'they who search after the Philosopher's Stone are by their own rules obliged to a strict & religious life. That study is fruitful of experiments.' He had become a kind of hermit, therefore, a hermit of the occult. He was intent upon understanding the secret structure of the world and the key to all learning. He was seeking the secret of the universe itself. It was the same force, and the same ambition, that inspired all of his endeavours and that led him eventually to the solutions of the *Principia Mathematica*.

There is one other aspect of his alchemical thought that bears a central role in his investigations. It was believed that, as a discipline, alchemy had its origins in remote antiquity, and that it had been practised by the magicians of Egypt and of Greece. It was sometimes claimed that Moses was the first alchemist. Newton himself was a firm believer in what was known as '*prisca sapientia*' or ancient wisdom, and even went so far as to claim that his mathematical researches were only rediscovering the lost principles of Pythagoras. He trusted in the knowledge of the ancients as an untapped source of great power that could be released into the modern age. Magical thought, or alchemical thought, was one way of conflating that ancient wisdom with modern

experimental techniques. John Maynard Keynes, one of the
first to read and reveal the content of Newton's unpublished
papers on alchemy, described him in a public lecture in 1946
as 'the last of the magicians, the last of the Babylonians and
Sumerians' who could look upon the worlds visible and
invisible with serene eyes. And indeed he has been afforded
the status of a magician, one who solved the enigma of the
universe and then revealed it to the initiated. We still live in
a Newtonian universe.

The idea of alchemical adepts, of unpublished texts and of
secret studies, powerfully appealed to his cloistered and
fugitive nature. When one chemical and alchemical experimenter, Robert Boyle, proposed a certain 'mercury' as an
agent of transformation Newton implored him not to
publish his results; they were 'an inlet to something more
noble, not to be communicated without immense dammage
to the world' if they were revealed. Newton was by
temperament isolated and secretive, unwilling to share his
knowledge, ever vulnerable and suspicious, hiding his work
with anagrams and conundrums. One of his contemporaries
at Cambridge described him as 'of the most fearful, cautious
and suspicious temper that I ever knew'. Alchemy was for
him the perfect lonely pursuit. So in secrecy and darkness he
sublimated, he dissolved, he distilled and he calcined.

## Chapter Five
# The professor

Yet, while secluded in his laboratory, he did not neglect his other studies. He could not afford to lose sight of the speculations that had brought him preferment. In 1669, after an interval of two years, he resumed his pursuit of mathematics or, rather, he was obliged to assert his authority in that discipline. Newton's tutor at Cambridge, Isaac Barrow, had received from a friend in London a copy of Nicolaus Mercator's *Logarithmotechnia*; this was a book in which Mercator, a German mathematician, outlined a more simple method of calculating logarithms. Newton had already accomplished this feat three years before, and had even gone beyond it, so he felt obliged to write down his own account in a treatise entitled *De Analysi per Aequationes Numeri Terminorum Infinitas* or 'On Analysis By Infinite Series'. He lent a copy of this text to Barrow, but refused to have it sent elsewhere or published. He relented to the extent that Barrow was allowed to send it to a colleague in London, but on no account was the work to be published. It is yet another indication of his cloistered and nervous temperament. It is as if he believed that, if he exposed anything of his to the world, he would be attacked.

On 29 October 1669, Newton was appointed the second Lucasian Professor of Mathematics, taking over that position from Isaac Barrow. He was still only twenty-six years old, eight years after he had first entered Cambridge as a freshman, and must rank as one of the youngest professors ever appointed in that university. Barrow himself was now on the friendliest terms with Newton, and appreciated his superlative gifts. He even asked Newton to edit his lectures on optics; Newton duly obliged without referring once to his own revolutionary experiments in that field. Barrow wrote to an acquaintance, John Collins, in London that 'he hath a very excellent genius to these things' by which he meant things mathematical. So when Barrow was appointed as chaplain to the reigning monarch, Charles II, he arranged that his professorial seat should be passed on to his young contemporary.

Newton's statutory duties included the stipulation that he lecture or expound each week, through the three terms, 'some part of Geometry, Astronomy, Geography, Optics, Statics, or some other Mathematical discipline'. He was fined forty shillings if he missed a lecture, and at the end of the academic year he had to deposit copies of his lectures in the university library.

It cannot be said that Newton was a good or a natural teacher. He had until that moment had only one pupil, St Leger Scroope, who has left no record of his tutorials under the supervision of a genius. Nor was he a natural lecturer. A later assistant remarked that 'so few went to hear Him, & fewer yet understood him, that oftimes he did in a manner, for want of Hearers, read to the Walls'. This was not in fact an altogether unusual experience for lecturers in seventeenth-century Cambridge, where study and discipline were equally lax. It cannot be said that lecturers were necessarily any better than students in that respect, and many of them ignored or circumvented their academic obligations. Newton himself was content to lecture during only one term out of the stipulated three.

His life in college had now acquired a settled status and routine. The Lucasian professorship afforded Newton a salary of one hundred pounds per year, as well as the other preferments and stipends of Trinity. His appointment gave him the time, and the freedom, to pursue his own researches. In this long period of his professorship he also acquired a reputation for eccentricity and absent-mindedness. When he attended Hall in college he was often so preoccupied with his calculations that he forgot to partake of anything, and the cloth was removed before he had eaten. He would make his way to the wrong church for divine service, or would wear his surplice at dinner. If he was entertaining friends and stepped into his study to obtain a bottle of wine, 'and a thought came into his head, he would sit down to paper and forget his friends'. He had a college servant, one Caverley, and a female servant known as 'Goodwif Powell', who were no doubt accustomed to his ways.

A later assistant, Humphrey Newton (see Chapter Seven), stated that 'he always kept close to his studies, very rarely went a visiting, & had as few Visiters'. The same assistant also wrote that 'I never knew him take any Recreation or Pastime, either in Riding out to take the Air, Walking, Bowling, or any other Exercise whatever, Thinking all Hours lost, that was not spent in his studies'. Here is the portrait of a secluded if not precisely misanthropic scholar, a pale devotee of sometime secret sciences. He is, in William

Blake's phrase, 'the virgin shrouded in snow', in the tradition of other melancholy and reclusive scholars such as Boyle and Evelyn.

It seems that he had one acquaintance, a chemist named Vigani, but he dropped him after Vigani had 'told a loose story about a Nun'. So the young professor was deeply religious as well as isolated. It is not perhaps a very attractive picture but it is the appropriate setting for one whose work was popularly regarded as being somehow 'superhuman' in its achievement. The stewards' books of Trinity also reveal that he scarcely left the university, except to make rare trips to his family home.

A month after being elected, however, he did make a second journey to London. Since he purchased alchemical books and instruments in the course of that visit, it is likely that he went there in order to make or continue his contacts with a secretive circle of adepts. He also took the opportunity of meeting Isaac Barrow's confidant, John Collins, who was at the centre of the mathematical discourse of the time. Collins recalled meeting him 'late upon a Saturday night at his Inne', when they conversed upon the nature of musical progressions. This heralded a more general correspondence between the two men. We can be sure that he did not pass any of his time in the capital pursuing frivolous amusements; he maintained a fierce concentration upon his work, to which everything else was subordinate.

On his return to Cambridge he prepared his first course of lectures as Lucasian Professor, on the subject of optics. He had told Collins that he was going to continue where Barrow had left off, but in that proposal he was being too modest. He announced in January 1670, to whatever audience of students was present, that 'I judge it will not be unacceptable if I bring the Principles of this Science to a more strict Examination, and subjoin, what I have discovered in these Matters, and found to be true to manifold Experience'. He was intending to reveal to his auditors, in other words, the nature of light that he had discovered in the course of his own prismatic experiments: white light contained an infinite number of colours, dependent upon their angle of refraction. It must have been rare indeed for any lecturer at Cambridge to have prepared himself so thoroughly for his task. The attendant students, if there were any at all, were not sufficiently impressed to leave any record of these occasions.

He may not have been particularly concerned by their lack of interest. When Collins wrote to him from London, requesting that some of his mathematical calculations be published, he agreed to the proposal 'soe it bee without my name to it. For I see not what there is desirable in publick esteeme, were I able to acquire & maintaine it. It would perhaps increase my acquaintance, the thing which I cheifly study to decline.' Here is the clear evidence of a man who wrapped himself in isolation and who savoured his singularity; his solitude was his carapace in which he could hide himself. His refusal to collaborate, and his difficulty in communicating his most profound ideas, are part of his essential character. At this late date, perhaps, we may exclaim with the Countess in *All's Well That Ends Well*, 'Now I see the mystery of your loneliness.'

In the spring of the following year he returned to Woolsthorpe. But he was still in correspondence with John Collins, who seems to have ensured that Newton did not lose interest in mathematics. He asked Newton to supervise the publication of a Latin textbook on algebra, and from time to time sent him what he considered to be interesting or significant books. When he sent Newton a copy of Giovanni Borelli's *De Motionibus*, however, he received a sharp riposte. Newton asked him to send no more volumes, since 'I shall take it for a great favour if in your letters you will onely inform mee of the names of the best of those bookes which newly come forth'. He did not wish to owe anyone an obligation.

He was still pursuing his mathematical research. He composed a tract entitled *Method of Fluxions and Infinite Series* in which he speculated on the introduction of infinitely small or 'indefinite' elements into equations. He had previously written a paper on rotational forces, but now he also began work on a treatise entitled *De gravitatione et aequipondio fluidorum,* but this account of fluid mechanics is characterised by Newton's strong emphasis on the constant presence and intervention of God within the natural world. His concern with the matters later raised in the *Principia* is seen to be compatible with his alchemical and theological interests. None of these papers was given to the public or even to the mathematical community. In 1671 he also enlarged his earlier paper, *De Analysi,* but this revision was also not published.

Collins was mightily impressed by the young professor's speculations. He had also become aware of Newton's suspicious and solitary temper. On the question of publication Collins told a friend that 'observing in him an unwillingness to impart, or at least an unwillingness to be at the paines of so doing, I desist, and doe not trouble him any more'. That was by far the best policy, and one that did bear unexpected fruit.

At the end of 1671, through the agency of Collins and of Barrow, Newton allowed his six-inch reflecting telescope to be examined by the members of the Royal Society. Barrow carried it from Cambridge to London, where the secretary of the society, Henry Oldenburg, arranged for the instrument to be exhibited at the society's premises in Gresham College on the corner of Bishopsgate. It proved to be a great success. It was carried in triumph to Whitehall, where Charles II was graciously pleased to accept it, and the Astronomer Royal described it to Collins as 'this prodigie of arte'.

Less than three weeks later Oldenburg wrote to Newton congratulating him on his excellent device and advising him that he had been nominated for election as a Fellow of the Royal Society. Newton replied in what were for him fulsome terms and concluded that 'I shall endeavour to testify my gratitude by communicating what my poore & solitary endeavours can effect towards the promoting your Philosophicall designes'. No one could have known, or guessed, just how extraordinary these poor and solitary endeavours would become. He was duly elected a Fellow on 11 January 1672, and remained attached to that institution for the rest of his life.

The Royal Society had been established in the rooms of an Oxford don twenty-four years before, but the Fellows had only decided to meet regularly in 1660. They were granted a royal charter in 1662. They excluded all questions of politics and religion, wisely enough after the disturbances of the revolution and the Restoration, and their motto became *Nullius in verba*: 'Nothing in words', or nothing on authority. They were interested in facts and in plain English; they were not interested in ideology, let alone 'enthusiasm'. They wished to proceed according to the precepts of practicality and pragmatism. The pursuit of science seemed to them to be the way of quieting sullen civil discord. It was in that sense a very English pursuit. They were in many respects a heterogeneous group of natural philosophers and experimenters, whose discussions and transactions were largely based upon speculation, observation and the 'just fancy that' school of enquiry.

Eight days after his election, no doubt encouraged by his recognition and his new status, Newton wrote to Oldenburg explaining that he was happy to communicate the theories of light that had prompted his construction of the telescope. He described it as 'a philosophical discovery', adding that in his judgement it was 'the oddest if not the most considerable detection which has hitherto been made in the operations of Nature'. A stunning declaration, evincing extreme if justified ambition and self-confidence, it was his claim that the discovery of the true constituents of light would act as a watershed in natural science. So on 6 February Newton sent Oldenburg his paper, now entitled 'Theory of Light and Colours', which was duly read to the Fellows of the society two days later. Among its propositions were those stating that 'Colours are . . . *Original and connate properties*, which in divers Rays are divers' and that 'Light is a confused aggregate of Rays indued with all sorts of Colours'. Here was new knowledge.

Oldenburg replied in great excitement, reporting that the young professor's revolutionary understanding of light and colour had been well received. Newton wrote back an equally enthusiastic reply deeming it 'a great privelege' to be received by 'so judicious & impartiall an Assembly' rather than misinterpreted by 'a prejudic't & censorious multitude'. He permitted Oldenburg to print his paper in the house journal of the Royal Society, the *Philosophical Transactions*. At that point Newton was enrolled among the European community of natural philosophers, and the anonymity he had once sought so assiduously was lost to him for ever.

# Chapter Six
# A secret faith

One of those who had listened to his paper in Gresham College, however, had not been so favourably impressed. Robert Hooke wrote to Newton praising the 'niceness and curiosity' of his experiments, and then proceeded to dismiss them in favour of his own preferred 'wave theory' of light. The criticism was perhaps all the more pointed since Hooke was curator of experiments at the Royal Society. Newton did not respond to him directly but sent a letter to Oldenburg in which he belittled Hooke's criticisms; he did not doubt that upon stricter testing his theory 'will bee found as certain a truth as I have asserted it'. Newton himself believed that light comprised particles or 'corpuscles', and remarked in passing that 'it can no longer be disputed . . . whether Light be a Body'. But he did not wish to press the point. He was intent upon his theory of heterogeneity.

Hooke himself was an inveterate experimenter and theorist, of whom his first biographer wrote that 'the fertility of his Invention . . . hurry'd him on, in the quest of new Entertainments, neglecting the former Discoveries'. He was in truth a child of the new age of experiment, an assiduous and voracious natural philosopher who took the whole world of inquiry as his province. Unlike Newton, however, he was convivial and gregarious; he was perfectly attuned to the coffee-house life of London in which artists and philosophers, poets and experimenters, might exchange conversation in places of resort such as the Grecian and the Rainbow. He did not take altogether kindly to the solemn and secretive Newton. He had great powers of intuition and 'Invention' but, again unlike Newton, he was not blessed with a mathematical or analytical mind. This did not prevent him, however, from claiming precedence over Newton in many areas of research. When Hooke stated that he had discovered the law of gravity before Newton, for example, Newton replied that 'I know he hath not geometry enough to do it'. The two contemporaries were bound to clash.

Hooke was joined in his criticism of Newton's optics by natural philosophers from the Continent, among them Huygens and the English Jesuits from the university foundation of Liège who were known for their scientific endeavours. One Jesuit had the temerity to refer to Newton's 'hypothesis', a description to which Newton took exception. He had proved his case, theoretically if not experimentally, by the most painstaking observation and by the most rigorous calculation: there was no hypothesis involved at all. If anyone wished to cast doubt about his conclusions, then they must do so by a process of experiment. That was, for him, how science worked. As in all of his speculations, Newton was at pains to arrive at what he considered to be mathematical truth. As he said in reply to Hooke, 'the science of colours becomes a speculation more proper for mathematicians than naturalists'. It can in fact be said that he mathematised nature. That is why opponents like William Blake, and the Romantic poets, conceived of him as the enemy of the imagination.

He was more disconcerted by the contemporary criticism than he allowed himself to appear. He believed that he had unveiled the 'most considerable' discovery in the history of science, only to be doubted and questioned by those who were manifestly inferior to himself. When Collins offered to publish his lectures on optics he demurred, 'finding already by that little use I have made of the Presse, that I shall not enjoy my former serene liberty till I have done with it'. He had lost that 'serene liberty' in the controversy over his paper in *Philosophical Transactions*. He had been enmired in the world, in questioning and in cross-questioning, in being forced to justify himself. Even his experimental accuracy had been questioned. Such things were intolerable to him.

He travelled to Woolsthorpe in the summer of 1672, perhaps hoping to clear his head of the controversy that had so unexpectedly and unhappily descended upon him. He returned there in the following spring. But these intervals did not soothe him. When he returned to Cambridge he wrote a letter to Henry Oldenburg, demanding that 'I may be put out from being any longer fellow of your R. Society. For though I honour that body, yet since I see I shall neither profit them, nor (by reason of this distance) can partake of the advantage of their Assemblies, I desire to withdraw.' Oldenburg then tried to placate Newton, and promised that he would no longer be liable to the quarterly payments for membership. The matter was allowed to drop. Newton never did leave the Royal Society.

But his response to the criticism from that institution was indicative of his response to the world. He took umbrage; he wished to withdraw before suffering any further notice or humiliation. Newton had complained to Collins at the same time that he had encountered 'rudeness'. His high self-esteem was matched only by his extreme sensitivity to attack. When Oldenburg wrote in a later missive that he should 'passe by the incongruities' offered to him by Hooke and others, Newton informed him that 'I intend to be no further solicitous about matters of Philosophy'. This sounds remarkably like petulance. He could not allow his belief in his rightness – even in his omnipotence, perhaps – to be in any way impugned. Yet the petulance is that of a child, and it may be related to his experiences as a child – in particular to the loss of his mother at an early age, when it must have seemed that the world conspired against him. He did not correspond with Oldenburg for eighteen months.

He had finished his course of lectures on optics and by the autumn of 1673 had begun a new set of lectures on arithmetic. These were difficult for the passing student inveigled into the lecture theatre, to the point of being incomprehensible, but he continued them for the next eleven years. The copies of ninety-seven lectures were, according to statute, eventually deposited in the University Library. His status in the college was now confirmed by his removal into a new and more luxurious set of chambers. These were at the front of the college, on the first floor, between the great gate and the chapel. There was a small garden, in which he used to stroll during the course of his meditations; there was also a stairway leading to a gallery, where he set up his reflecting telescope. There seems also to have been a small shed or annexe, in which he established his laboratory. Wickins had moved with him, and no doubt still acted as his assistant.

Here he continued with what one contemporary

described as his 'Chimicall Studies'. He had already explained to Oldenburg that he was engaged in 'some other subjects' and 'business of my own which at present almost take up my time & thoughts', which can be construed as meaning his recondite alchemical experiments. At the same time he became preoccupied with another subject that was deeply implicated in his study of alchemy and in his exploration of the wisdom of the ancients. He became a student of the Scriptures. On the back of a draft of the letter he composed to Oldenburg, in which he stated his intention to leave off 'Philosophy', he jotted down material concerning the prophecies of the Old and New Testaments.

In particular he became intent upon the prophecies of Daniel and the Book of Revelation written by John the Divine. He was in search of eternal truth. There was for him no necessary disjunction between science and theology. They were part of the same pursuit. Theology and science were, equally, avenues to God. They were the keys to true knowledge of the universe. He was a philosopher in the ancient sense, a seeker after wisdom. In his earlier treatise, *De gravitatione*, he had suggested that 'the analogy between the Divine faculties and our own is greater than has formally been perceived by Philosophers'. He wished to bring himself closer to the divine.

Newton's study of the Old Testament was, as was to be expected, rigorous and thorough. He had more than thirty different versions or translations of the Bible. He learned Hebrew in order to study the original texts of the prophets. He began a notebook in which he schematised his study, with headings such as 'Incarnatio' and 'Deus Pater'. He amassed a huge library of biblical and patristic literature. He read all the authorities of previous centuries, and assimilated the most modern texts of seventeenth-century theology in his desire for true knowledge. He wished to become the master of his subject, as he had previously become the master of optics and the master of mathematics. At his death he left a manuscript on biblical matters, incomplete, of some 850 pages as well as a mass of assorted papers and notes.

In particular he became preoccupied with a dispute of the fourth century, during the course of which he determined that the true faith – Protestantism, as he conceived it – had taken a perverse and highly injurious turn. The great controversy was between Arius and Athanasius. Athanasius propounded what had then become the orthodox doctrine of the Trinity, in which Christ is seen as equal or 'consubstantial' with God. Arius denied the doctrine of the Trinity by denying that Christ was of the same substance as God. The views of Athanasius were accepted at the Council of Nicaea in 325, and of course became a part of the Nicene Creed.

But in the course of his intense study of the biblical texts Newton concluded that Athanasius had perpetrated a fraud. He had interpolated key words into the sacred Scriptures to support his argument that Christ was God. In that endeavour he had been supported by the Church of Rome, and from that corruption of the texts had sprung the general corruption of the Christian Church itself. The purity and faith of the early Church had been destroyed by superstitious zealots who were intent upon bowing down before the illusion of the Trinity or Three In One. His mathematical, as well as his spiritual, creed directly opposed their position. In his support of Arius Newton was proclaiming that the priests and bishops of the Church were practising idolatry in their worship of Christ. Newton discovered, in the words of a fellow Arian, 'that what has been long called Arianism is no other than Old uncorrupt Christianity; and that Athanasius was the grand and very wicked Instrument of that Change'. In his notebook Newton declared that 'the Father is God of the Son'.

Newton also believed that the true religion was derived from the sons of Noah, and had been transmitted by Abraham, Isaac and Moses. Pythagoras was a convert to this religion, and passed it on to his own disciples. Christ was a witness to that primitive faith in his simple commandments to love God and to love one's neighbour. In a later document Newton declared that we must worship 'the only invisible God' and venerate the 'one mediator between God & man the man Christ Jesus'. At the peril of our souls 'we must not pray to two Gods'. We must not worship Christ. Christ had been filled with divine spirit, but he was not God.

The fact was that, in the mid-seventeenth century, Arianism was still considered to be a dangerous heresy. If Newton had admitted his faith he would have been stripped of his university appointments, as were other and less cautious Arians. So he did not discuss these matters openly. He reserved his theological conversations for fellow believers. The full scale of his religious heterodoxy was not revealed until after his death, and even then the knowledge of it was suppressed by those scholars who believed that the father of English science must be above suspicion. To all outward appearances he remained a firm and orthodox member of the Church of England, leaning somewhat to the dissenting or radical tradition within that Church. But nothing more.

There were other, and perhaps more curious, aspects of Newton's secret faith. He knew by heart the words of the angels to St John, 'Rise, and measure the temple of God . . .' He took the instruction literally, and from ancient documents measured the dimensions of the Temple of Solomon. Newton believed that Solomon, the son of David and great king of the Jews, was 'the greatest Philosopher in the world'. He believed that Solomon had imbibed the wisdom of the ancients and that, in the design of his temple, he had incorporated the pattern of the universe. The sacred fire at the centre of the temple was the fire of the sun. It was an interesting theory, but one that he followed up by constructing a detailed plan of the edifice. In this image we may see the obsessiveness no less than the beauty of Newton's mind, creating intricate shapes in an abstract world of thought and imagination. In one enlightening passage Newton comments upon the language of dreams in the Old Testament. It is perhaps appropriate that the discoverer of universal gravity was also an analyst of dreams.

He was intent, too, upon the nature of biblical prophecy. He pored over the prophets, tracing the path of their utterance through symbols and hieroglyphs. He believed that in their words could be found hidden truths concerning the future history of the world. He drew up a catalogue of seventy inspired men, noting down the details of their lives and writings. He compiled a dictionary of world events that were deemed to match their prophecies. And he wrote an essay entitled 'The Proof' in which he maintained the authenticity and accuracy of the prophets' words. The eleventh horn of the Beast of Revelation, for example, was the Church of Rome.

He also devised a chronology for the future as well as the past. This was a continuation of his work on the prophecies, and was characterised by the same imposition of formulae and rules of interpretation. In 1944 would end 'the tribulation of the Jews' – he was 'out' by one year – and in 2370 would begin a thousand years of peace. We may seem to be a world away from the optical experiments and the mathematical calculations of his public work but all his activities evince the rapt contemplation of the magus poring over the universe. Yet even in these arcane studies he had not lost his empirical mastery. He calculated one aspect of his chronology by measuring the life-cycle of the locust. He dated the expedition of the Argonauts, to find the Golden Fleece, by measuring the solstice and the equinox. He wanted to clarify, and therefore bring within his control, the machinery of the universe.

In theory and in practice, therefore, his scientific and religious studies (if we can even make that distinction) were connected. As he wrote of the apostle John: 'I have that honour for him, as to believe that he wrote good sense; and therefore take that sense to be *his*, which is the best.' The same clear-eyed observation dictated his practice in the laboratory, even when he was labouring over the secret arts of alchemy. There seems no doubt that he believed his destiny to lie in the discovery and interpretation of the works of God. His discovery of universal gravity was further proof of the divine plan, and he declared that God was everywhere within his creation. He was the '*Lord God pancrator* or *Universal Ruler*', and Isaac Newton was his servant. On one occasion the philosopher and theologian, Henry More, had conversed with him at Cambridge on the Apocalypse. More recollected that Newton, ordinarily 'melancholy and thoughtfull', was by the end of their discussion 'mighty lightsome and chearfull' and 'in a maner transported'. This description of his 'transport' suggests the elevation of mind and feeling that accompanied his explorations of the divine world.

Yet his conscience did not cease to trouble him. Newton could not retain his fellowship at Trinity indefinitely without taking holy orders. This would mean, in particular, that he would have to subscribe to the doctrine of the Holy Trinity. This he could not do. So he travelled to London in February 1675, with a petition to Charles II asking that he be excused taking holy orders while he was still Professor of Mathematics. It was a technical excuse designed to cover a momentous difficulty. He waited in the capital for a month, until the petition was granted. The king declared that he wished 'to give all just encouragement to learned men who are & shall be elected to the said professorship'. It is a signal instance of the new regard now held in England for mathematics and natural philosophy.

While in London he relented his disavowal of the Royal Society, and attended two of its meetings. He was happily surprised by the reception he received, having previously mistaken criticism for hostility, and even agreed that a series of experiments should be conducted to confirm his theories of the prism. This was in fact going to be a much delayed project, and it was not until the winter of the year that he wrote to Oldenburg on the form and nature of the trials. He told Oldenburg that he had had some intention of writing another paper on colours but had changed his mind, finding it 'yet against the grain to put pen to paper any more on the subject'. However, 'I have one discourse by me of that subject written when I first sent my letters to you about colours . . .'

He also sent an explanation of his theories, which he entitled 'Hypothesis Concerning the Properties of Light'. Newton's 'aetherial hypothesis', as it has become known, is chiefly remarkable for his controversial speculation that 'perhaps the whole frame of nature may be nothing but various contextures of some certain aethereal spirits, or vapours, condensed as it were by precipitation'. There is a suggestion here of the influence of his alchemical experiments, and it represents his first foray into what might literally be described as the cosmic world. It should be added, however, that he refused to permit this paper to be published. Instead it was read and discussed by the members of the Royal Society at four separate meetings, at the end of 1675 and the beginning of 1676, although he was not present at any of them. He had already informed Oldenburg that he did not feel 'oblig'd to answer objections against this script' since 'I desire to decline being involved in such troublesome & insignificant Disputes'. It is clear that he was acutely aware of his own superiority.

At the same time he had entered into correspondence with Robert Hooke, whom he had deemed to be an opponent of his work. They had met when Newton had attended the meetings of the Royal Society earlier in the year, and seemed to have been able to overcome their differences. Newton also laboured under the mistaken belief that Hooke now accepted his theory of colours, a mistake perhaps due to the elaborate politeness and formality with which seventeenth-century gentlemen customarily comported themselves.

After hearing Newton's paper read aloud in Gresham College at the end of 1675, however, Hooke stated that the younger man was indebted to the researches in his own *Micrographia*. In turn Newton replied that Hooke had been unable to master the calculations involved in the optical experiments. It could have been the beginning of an unpleasant war of words over priority, the kind of dispute in which Newton excelled in later life, but Hooke mollified his younger rival with weasel words. Hooke confessed that he himself had no time or leisure to complete his early work, and also pleaded 'abilities much inferior to yours'.

It was the kind of surrender Newton always demanded from his opponents. He replied gracefully, congratulating Hooke for his 'generous freedom' and adding that 'you have done what becomes a true philosophical spirit'. He is happy to continue a private correspondence with Hooke (which in fact he did not) on the assumption that 'consultation' is better than 'contention'. In the course of this letter Newton made his famous remark that 'If I have seen farther, it is by standing on the shoulders of giants'. It may be unfair to note here that Hooke was of small and somewhat crooked stature.

They agreed, in any case, that the crucial experiment on Newton's theories of the prism should take place at Gresham College in the spring of 1676. The results were all that Newton could have wished, and his speculations were thoroughly justified by the result of the public experiments. There was still some criticism from various offended parties, but the major controversy had now been settled in Newton's favour. As Oldenburg put it in his minutes of this significant meeting, the experiment 'was tried before the Society, according to MR NEWTON'S directions, and succeeded as he all along had asserted it would do'.

In the following year, however, Newton's first great supporter, Isaac Barrow, died. He was followed by Henry Oldenburg, Newton's principal defender in the Royal Society. When Oldenburg was succeeded in his position as secretary by Robert Hooke, it seems that Newton felt himself to be once more threatened and isolated. He withdrew himself from the business of the Royal Society. He had also been reluctant to enter into a correspondence over mathematics with the celebrated German natural philosopher, Leibniz, and confessed that 'if I get free of this business I will resolutely bid adew to it eternally'. He did in fact leave mathematics alone for a further seven years. He immured himself in his study and laboratory at Trinity College, where he laboured over the secrets of the world. He would not emerge into the public sphere for another six years.

## Chapter Seven
# A taste of fire

In his Cambridge enclave and retreat he walked in his garden where, according to one of his assistants, he could not bear the sight or presence of any weeds. This was part of his drive towards order, neatness, and perfection. He kept a box filled with guineas by the window, as a deliberate test of the honesty of those who worked for him. He liked eating roast apples in winter, and one of his most unlikely letters is concerned with the proper making of cider. He gave money to the new library in his college, now universally known as the Wren Library, and was consulted by other colleges on various technical matters. He was in other words the image of a respectable and reclusive professor.

Some of his studies may have gone awry, however, since there are persistent reports that in 1677 there was a fire in his Cambridge rooms. His relative, John Conduitt, left a note concerning Newton's memory of this fire. 'When he was in the midst of his discoveries', he wrote, 'he left a candle on his table amongst his papers & went down to the Bowling green & meeting somebody that diverted him from returning as he intended the candle sett fire to his papers.' Newton recalled that these 'papers' were concerned with optics and with mathematics that 'he could never recover'.

There are other reports of the fire (though these, in fact,

may relate to other fires). In one of them Newton returned from the college chapel to find a book of his experiments incinerated, at which he became so agitated 'every one thought he would have run mad, he was so troubled thereat that he was not himself for a Month thereafter'. One of his assistants also told Stukeley that, on this or another occasion, 'a piece of chymistry, explaining the principles of that mysterious art upon experimental and mathematical proofs' took fire in his laboratory; after the incident Newton averred that 'he would never undertake that work again'. It is hard not to relate these incidents of fire to his work in the alchemical laboratory, where many experiments needed the presence of perpetual flame. It is unlikely, however, that anything of value was lost for ever; Newton was so meticulous in his notes, and so intent upon redrafting and revision, that he would have been able to reconstruct any of his work.

The report of his extreme agitation is not far removed from later and more grotesque accounts of his 'madness'. It is a commonplace that genius is near allied to insanity, and it is the fate of many men of imagination to be reported to be out of their wits. This is the charge that mediocrity hurls against great achievers. Yet it is clear enough that Newton did often possess a very troubled mind. He was engaged in a brief correspondence in 1678, for example, with an experimenter who doubted Newton's theories of light and colour. To this gentleman, in the spring of 1678, he asked: 'Am I bound to satisfy you? It seems you thought it not enough to propound Objections unless you might insult over me for my inability to answer them all . . .' There was much more in the same vein that, given the generally formal and polite tone of scholarly address in the seventeenth century, was little short of an explosion of fury. When John Aubrey informed Newton that he had another letter on the same subject Newton replied briefly: 'Pray forbear to send me anything more of that nature.'

He could not bear to be criticised or questioned in any way. He told another natural philosopher that he had been 'persecuted with discussions arising out of my theory of light', and in the choice of 'persecuted' there is a suspicion of what we might call paranoia. The fervent anxiety and intensity of his nature are not in doubt. He was one of those who could never rest in life. His mental endurance, his ability to keep a problem in his mind for days or months on end, is probably without parallel.

Yet he also had a need for enclosure, for a carapace in which he could conceal himself from the world. This may be related to the apparent fact that he spent a childhood without close familial love, his father and mother being absent from his earliest years, and as a result he craved security and order. He needed to be safe and inviolate. The deep need for order may have propelled his search for law and system in the universe, but it certainly made him vulnerable to any form of attack. When he felt himself under threat, he lashed out. Maynard Keynes, in his 1947 lecture at Cambridge, described his 'profound shrinking from the world, a paralyzing fear of exposing his thoughts', as if the skin were being ripped from his body in the process. That is why throughout his life he remained isolated, reclusive and remote.

That isolation increased when, in the spring of 1679, his mother died. In May she had nursed her son, Benjamin, through what was called a 'malignant feaver'; he survived, but she contracted the distemper. When Newton was informed of her deteriorating condition he travelled up to Lincolnshire at once and, according to John Conduitt, 'sate up whole nights with her'. At the same time he 'gave her all her Physick himself, dressed all her blisters with his own hands, & made use of that manual dexterity for which he was so remarkable'. This account tends to nullify suggestions that Newton felt some primal anger against his mother for leaving him in the care of his grandmother, and modifies the other picture of him as passionless and uncaring.

But his skill was not enough. Hannah Smith died at the end of May, and in the following week was buried in the churchyard of the neighbouring village of Colsterworth close to the grave of Newton's father. In her will, apart from incidental bequests, she left him all her lands and goods. He remained in his family home for some six months, during which period he looked after his inheritance. He was now a man of substance. He had dealings with tenants, and may even have supervised the autumn harvest. He was also relentless in chasing up outstanding debtors. To one debtor he wrote, 'I shall only tell you in general that I understand your way & therefore sue you. And if you intend to be put to no further charges you must be quick in payment for I intend to loos no time.'

On the day after his return to Cambridge on 27 November, he wrote to Robert Hooke explaining why he had not communicated with him as he had once promised. He described himself as suffering from 'short sightedness and tenderness of health', but this may have been in part an excuse or a sign of his general tendency towards hypochondria. He acknowledged somewhat circumspectly that 'I have been this last half year in Lincolnshire cumbred with concerns among my relations'. His attention to 'Countrey affairs' had pre-empted any 'Philosophical meditations'. But then he declined to take any further part in such enquiries, 'having thus shook hands with Philosophy' and said farewell. He also made the homely analogy about his lack of interest 'as one tradesman uses to be about another man's trade', suggesting that natural philosophy was a business like any other.

But he was not being altogether truthful about his loss of interest. In the earlier part of the year he had written to Boyle on 'a certain secret principle in nature' which would explain the cohesion of certain substances. In the same period he had also written to Locke on the physical principles of gravitation. This was not someone who had abandoned philosophy. In the letter to Hooke he explained that he was in the process of building a new reflecting telescope and expounded at some length what he called 'a fancy of my own about discovering the earth's diurnal motion'. So he had not given up on speculation or on observation; he just did not want to be bothered by questions or criticisms.

The matter of 'the earth's diurnal motion' concerned the path of a heavy body falling to the earth, which in his letter Newton deemed to be a spiral. Robert Hooke, ever watchful for flaws in the reasoning of the celebrated Newton, discovered a mistake in his argument. But he did not keep his reservation for private correspondence. He announced Newton's error to the assembled Fellows of the Royal Society. A falling object would act in the manner of an orbiting planet and form, not a spiral as Newton would have it, but 'perhaps an ellipse'.

Newton had in fact made an uncharacteristic mistake, and the knowledge of this slip distressed him. If he was not perfect, and therefore not inviolable, he was not himself. But, equally importantly, the public nature of this correction infuriated Newton. Hooke had promised him in his original letter that any material Newton sent would be 'no otherwise further imparted or disposed of than you yourself shall describe'; but he had flagrantly broken that promise in order to expose Newton to the derision of his contemporaries.

The response of Newton was predictable. He dealt with Hooke's next letter, which pointed out his mistake, briefly and coldly. He then refused to answer any of Hooke's subsequent letters, and indeed wrote to no one for a year. At a later date he described Hooke as one incapable of mathematical calculation who 'does nothing but pretend & grasp at all things'. Thirty years later, with Hooke long dead, he still considered his scientific rival to be his personal enemy. But his mistake led him to work through the problems of dynamics, and apply his mind to matters of orbital mechanics in which he claimed to have lost interest. In particular he seems to have calculated to his satisfaction that an elliptical orbit was consonant with a central force which diminished with distance. Out of the debacle with Hooke, in fact, came the seeds of the *Principia*. It is also significant that in this period he interested himself once more in the principles of classical geometry.

His curiosity in cosmological matters was quickened by a strange apparition in the night sky of November 1680. It appeared just before sunrise, and eventually vanished in the

direction of the sun. Then in December another comet
emerged early one evening, moving away from the sun. This
was a larger phenomenon, with a glowing 'tail' four times as
broad as the sun. The Astronomer Royal, John Flamsteed,
wrote to a friend that 'I believe scarce a larger hath ever been
seen'.

Newton picked up the trail of the December phenomenon. He began keeping information on its progress,
watching its 'tail' with particular interest. At first he used a
concave lens placed against his eye, to correct his short-
sightedness, but as the comet grew less distinct he observed
it with a telescope. He watched until it finally disappeared
on 9 March. He was so intensely interested in the subject
that he gathered reports from astronomers in Europe, and
even received news of the sighting from the banks of the
Patuxent river in Maryland. One of his contemporaries from
Grantham, who had emigrated to that area, wrote to him
that the comet had a 'form like a sword streaming from the
horizon'. The experience of tracking it did in fact impel him
to begin constructing a much larger reflecting telescope, but
this he did not finish. The experience of the comet did, however, furnish him with much material for contemplation.

Most observers were convinced that they had witnessed
two comets but John Flamsteed firmly believed that the two
appearances were of the same body moving around the sun.
He communicated his theory to Newton, who initially
rejected the idea – or, rather, rejected Flamsteed's speculations about the magnetic poles of the sun. It is clear
enough that he had not yet devised his theory of universal
gravitation. But he began to plot possible trajectories, and
to test the theory that this comet had indeed moved in an

elliptical orbit rather than being a singular and random irruption from space. He pondered the implications, and they would bear fruit within a few years. In time he also devised a theory concerning the purpose of comets. In the first edition of *Principia Mathematica* he asserted his belief that the matter glowing in the 'tail' of comets furnished vital material for the replenishment of life on earth. He also suggested, in the second edition of the treatise, that comets themselves might refuel the sun and stars. These may be considered to be instances of the occult nature of Newton's scientific thought. His notebooks testify that he was still intent upon alchemical experiment.

In the interim he lost his close colleague and assistant, John Wickins, who had shared chambers with him since their days as undergraduates. They had been close companions for twenty years but in 1683 Wickins resigned his own fellowship at Trinity in order to take up a clerical 'living' in the country. They corresponded only once after his departure, with an exchange of brief letters that do not imply any great intimacy. It is possible that they had simply grown incompatible, or that some disagreement had arisen between them. It also seems that Newton sent numerous Bibles to Wickins, to be distributed to the poor people of his neighbourhood, which shows a continuation of religious sympathy at least.

Wickins's place was taken by a promising young scholar from the school at Grantham which Newton had once attended. The young man's name, Humphrey Newton, has suggested to some that he was a relative; but no evidence of that has ever been discovered, and the younger Newton himself made no claim of that kind. Since Newton was not a rare surname in Lincolnshire, it seems to have been simple coincidence. He acted as Newton's assistant and amanuensis for five years, and in that capacity had ample scope for observation of his employer. In a later memoir he described him as 'very meek, sedate & humble, never seemingly angry, of profound thought, his countenance mild, pleasant and comely'. This is not the impression that others gathered of Newton, and it seems likely that Humphrey Newton was happy to present an idealised portrait.

He added that his employer ate and drank sparingly, was peculiarly absent-minded, and when attending to his studies did not go to bed until two or three in the morning. It seems that he slept in his clothes and, if not corrected, would go out into the street with his stockings unfastened and his hair uncombed. The fire was continually burning in his laboratory, although Humphrey Newton declares that 'what his aim might be I was not able to penetrate into', which suggests a measure of prudent reticence on his part. He also notes that he had only known Newton laugh once in five years, on an occasion when he was asked what was the use of studying Euclid. The exact meaning of this laugh is not at all clear. In this period, when he was touching forty, his hair was already grey. He blamed this condition on the prolonged contact with quicksilver (mercury) in his experiments, although others might ascribe it to fatigue and labour.

Humphrey Newton's most significant task was to transcribe the great work of Newton's career. 'I copied out,' he later recalled, 'before it went to the Press, that stupendous work.' He was of course referring to *Principia Mathematica*, the 'stupendous' achievement that had its genesis in the year after his appointment as secretary. In the summer of 1684 the astronomer and natural philosopher, Edmond Halley, travelled down to Cambridge in order to ask Newton what turned out to be a momentous question. The answer changed the whole nature of Newton's life.

## Chapter Eight
# Eureka!

In the early months of 1684, at a meeting of the Royal Society, Halley had been engaged in a scientific discussion with Christopher Wren and Robert Hooke on the dynamics of planetary motion. The question Halley put to them was at the time a significant one – could the force that kept the planets moving around the sun decrease as an inverse square of its distance? When Halley put the question to them, both Wren and Hooke burst out laughing. The 'inverse square law' was not a new concept at all. Hooke declared that it was the principle upon which all the laws of celestial motion were established, while Wren confessed that for some time he had been intent upon proving the hypothesis without being able to do so. Hooke then promised to provide a proof of his own within two months, but in the end failed to achieve it.

So Halley's thoughts wandered to the professor of mathematics in Trinity College. If anyone could provide convincing demonstration of the law, then it would be Isaac Newton. He considered writing to him, but he had been informed of Newton's cloistered existence. So he simply took a coach down to Cambridge, and braved the mathematician in his den. They soon became amicable enough for Halley to put the question. He asked Newton about the curve that would be described by the planets around the sun 'supposing the force of attraction towards the Sun to be reciprocal to the square of their distance from it'. Newton replied immediately that it would be an ellipse. Halley 'struck with joy and amazement', according to the memoirist who took down his recollections, asked him how he knew this. 'Why,' Newton replied, 'I have calculated it.' It was the first time that anyone had been able to achieve this feat. Halley asked him if he might see these calculations, and Newton rummaged through his papers. He could not find the precise notes, he told Halley, but he promised to redo them and send them on.

His innate prudence had held him back. He would renew his calculations, to guard them against error, before he gave them to Halley and to the world. He did in fact locate some mistakes in his original work but, inspired by Halley's enthusiasm, he worked upon the project with his usual unremitting zeal and attention. By November he had completed a short treatise of nine pages, entitled *De motu corporum in gyrum* – 'On the motion of bodies in an orbit'. As soon as he saw the paper Halley recognised its importance. For the first time the orbits of the planets had been deciphered and, more significantly, proved mathematically. He lost no time in returning to Cambridge, where he conferred with Newton on the best means of conveying these matters to the larger world.

But, as Halley put it later, he was the Ulysses who produced the Achilles. Newton did not stop at *De motu*, but went on to formulate a much larger theory. He wrote to Flamsteed asking for more data on the movement of the stars. He was concerned with tiny fluctuations in the movement of Saturn as well as requiring precise tables of the tides.

The whole of the known universe came under his scrutiny. 'Now I am upon this subject,' he told Flamsteed, 'I would gladly know the bottom of it before I publish my papers.' And so, for more than two years, he remained virtually in seclusion in order to complete his calculations. He made two short visits to Lincolnshire in the spring and early summer of 1685 but, apart from those absences, he remained for two and a half years in his college without respite.

Humphrey Newton recalled of this period that 'he has sometimes taken a Turn or two [in his garden], has made a sudden stand, turn'd himself about, run up the Stairs, like another Archimedes, with a eureka, fall to write on his Desk standing, without giving himself the Leasure to draw a Chair to sit down in'. He would forget to eat and, when reminded that he had left his food untouched, would exclaim, 'Have I!' before eating a little while still standing. He never bothered to sit down for his meals. This is the portrait of a man in the grip of an inspiration, or an obsession, that would never let him rest. He was on the verge of the greatest scientific discovery of the modern era.

The treatise *De motu* had been sent to Halley in November, and within the space of the next eighteen months Newton managed to complete the 550 pages of the treatise that would confer upon him the acclamation of the world. As Newton put it in a brief memorandum: 'The Book of the Principles was writ in about 17 or 18 months, whereof about two months were taken up with journeys, & the MS was sent to the RS in the Spring of 1686.' The 'Book of the Principles' was of course *Philosophiae Naturalis Principia Mathematica*, or 'The Mathematical Principles of Natural Philosophy'.

Newton had already calculated that the movement of the

planets around the sun, and the movement of the moon around the earth, were governed by the 'inverse square law'. But he wished to proceed further and create a general theory of celestial dynamics. In the original treatise upon which *Principia Mathematica* was based, *De motu*, there was no theory of universal gravitation; nor was there any description of what have since become known as Newton's three laws of motion. He had been concerned only with the inverse square law as it applied to the planets and to the comets. But his request to Flamsteed about the precise movement of the tides, in the Thames estuary, suggests that he was extending his theories of gravitation much further.

The three laws of motion are the foundation of his theory or, more pertinently, are the cornerstone of the universe itself. The first law states that every body 'continues in its state of rest, or of uniform motion in a right line' unless it is affected by an external force. The second law declares that this change of motion or direction is in proportion to the external force, 'and is made in the direction' of the straight line 'in which that force is impressed'. These laws were not in themselves particularly revelatory but then Newton added a further law to the effect that 'to every action there is always opposed an equal reaction'. He refined this by stating that 'the mutual action of two bodies upon each other are always equal, and directed to contrary parts'. This concept was something of an intellectual puzzle to his first readers, since it could not easily be visualised or demonstrated. So he produced a homely analogy, drawn from his days on the farm in Lincolnshire. If a horse draws a great stone with a rope, the horse will be drawn back to the stone as much as the stone is drawn towards the horse. The visible movement will proceed in the direction of the greater mass.

In the process of enlarging and refining his original perceptions Newton drew a novel and indeed revolutionary distinction between 'mass' and 'weight'; they were proportional, but not equivalent. Mass was the product of density and volume, whereas weight varies in various locations. Newton in fact introduced the concept of mass to the world, where it has remained ever since. He also introduced the term 'centripetal' as a key element in his theory of universal gravitation, by which he meant that in certain circumstances a body would naturally be attracted to the centre of another body. This was of course one of the principles of his new theory of gravitation.

Within a year the nine-page tract had grown to ten times the length, and had been divided into two books, *De motu corporum* and *De mundi systemate*; the first book dealt with the mathematics of orbital movement, and the second with a more general description of what he called in the preface 'rational mechanics'. But then he changed the plan of composition. He turned this second book into the third book of the finished work, and added a new second book on pendulums, wave motion, and, most importantly, fluid mechanics and fluid resistance. Much of the third book, then, was revised from his earlier work.

In this third book he outlined a set of principles or 'regulae' for the pursuit of natural philosophy. He then recounted the mathematics of the first book as a key to the understanding of universal gravity, which is followed by his theory of tides, his theories upon the motion of the moon, and finally his theory upon the motion of comets. As he explained in the course of this narrative, 'I derive from the celestial phenomena the forces of gravity with which bodies tend to the sun and the several planets. Then, from these forces, by other propositions which are also mathematical, I deduce the motions of the planets, the comets, the moon and the sea.'

It was an astounding achievement. Newton had announced his revolutionary principle of universal gravitation. The universe was interdependent, every part of it bound together by one force that could be mathematically promulgated and understood. He contrived the mathematics that would account for the force that keeps a body in orbit, and also for the curvilinear path that such a body would follow. This was the revelation. He had mathematicised the cosmos. He had made it amenable to human laws. In that enterprise he progressed according to an apparently simple formula, to the effect that 'Nature is exceeding simple and conformable to herself'. It was not a chaos, a bewildering mix of atoms and forces, but an explicable whole. No scientific treatise before had proceeded so diligently upon the basis of evidence; no scientific work had depended so exclusively upon observation and experiment. 'He that works with less accuracy is an imperfect mechanic,' he wrote in his preface, 'and if any would work with perfect accuracy, he would be the most perfect mechanic of all.' That is the description that has of course been given of Newton himself.

In the popular mind it is sometimes believed that somehow he was the first to discover or 'invent' gravity. But that is not the case. Copernicus and Kepler had already speculated about gravitational attraction. The force of Newton's originality lay in the fact that he demonstrated it mathematically, and proved that it was a universal force. No one before had proven beyond doubt, for example, that the tides were affected by the sun and the moon. That was Newton's discovery. He had proved the effects of invisible action at a distance, which before had been considered merely an occult fancy. He had also demonstrated that the forces of celestial and terrestrial motion were part of the same system. Every speck of matter in the universe was guided by the principles he had revealed. He was more than simply a perfect mechanic. He eventually became a magus. 'It is now established', he declared, 'that this force is gravity, and therefore we shall call it gravity from now on.'

The frontispiece of the first edition capitalised the words PHILOSOPHIAE and PRINCIPIA, and there seems no doubt that Newton was setting his own work against the treatise by Descartes entitled *Principiae Philosophiae*. He had called his own work *Principia Mathematica*, a wholly mathematical response to what he considered to be the errant hypothesising of the French philosopher. He was intent upon exposing the fallacies of Cartesian philosophy, with its conception of a mechanical universe and in particular its doctrine of 'vortices' or swirling pools of aetherial matter. Descartes believed that there was no gap or void in the universe and that it was filled with this invisible and amorphous matter. Newton did not agree. As the Dutch mathematician, Huygens, put it, 'Vortices have been destroyed by Newton'.

He had written the book in Latin so that it could be studied by the European community of natural philosophers. He admitted that he also rendered the subject more complex, and mathematically more advanced, in order to

ward off the prying eyes of the vulgar. There is an element
of alchemical secrecy and mystery about such a procedure,
but he also wanted to deter critics. As one acquaintance
reported, 'to avoid being baited by little Smatterers in
Mathematicks, he told me, he designedly made his Principia
abstruse'. He had achieved this goal with such success that
*Principia Mathematica* is still considered to be a great
challenge to students.

He called it in the preface to the first edition a study in
'rational mechanics'. But it would be quite wrong to state
that thereby the universe became fixed and immutable.
Newton recognised that the complexity of his vision, in
which every planet and star affected every other, added an
element of uncertainty in all calculations of relative motion.
As he said himself, 'it would exceed the force of human wit
to consider so many causes of motion at the same time'.

It has often been suggested that Newton would not have
been able to envision his theory of universal gravity (for
vision it certainly was) without the benefit of his alchemical
researches. It is certainly true that the concept of an invisible
force, acting between material particles, is one that he could
have derived from the textbooks of the adepts. The
alchemical pursuit was itself based upon the notion of a secret
principle that animated the material world, and the theory of
gravity can be seen as an aspect of those speculations. Yet of
course there is no mention of such matters in the *Principia*
itself. In his book he insisted upon the plain mathematics of
his discoveries. He asserted that 'natural philosophy' should
not 'be founded . . . on metaphysical opinions', and that its
conclusions can only 'be proved by Experiments'.

This may have been part of his obsessive need to conceal

his own predilection for alchemical concepts, as well as his own somewhat unorthodox religious and theological opinions, but as a result he helped to create the modern idea of science and the scientist – although the terms were not introduced until 1833. It is a great irony that Newton himself did not present the single-minded and rational image of the laboratory scientist; but that is the image of science that he almost single-handedly foisted upon the world.

In *Principia* he also created the scientific style, a deliberately neutral and plain prose borrowed from Locke but replete with numbers and figures that baffle those without mathematical training. There are no rhetorical flourishes, and scarcely even any adjectives. One example in the most recent translation from the Latin will mercifully pass for many – 'Therefore, as area PIGR decreases uniformly by the subtraction of the given moments, area Y increases in the ratio of PIGR – Y, and area Z increases . . .'. He was not writing a literary monument, however, but a textbook for the learned. He was very precise. He would write a phrase and then cross it out, substituting another word; then he would add a qualifying phrase; he would add and delete passages in the process of revision.

He was not averse, however, to fudging or doctoring figures so that he might pretend to a higher degree of accuracy than he had actually accomplished. In certain questions of gravity and velocity, he fixed his calculations in order to claim an exactitude of one part in three thousand. No one of course was in a position to check his figures properly, and so he got away with it. It suggests that Newton's vanity and desire to impress were still part of his bearing in the world.

# Chapter Nine
# The great work

By the spring of 1686 Newton had completed a large part of the manuscript, and sent it on to Edmond Halley who was pushing him towards publication. Halley thanked him for the 'incomparable treatise' but then went on to raise a very delicate topic. 'Mr Hook', he told Newton, 'has some pretensions upon the invention of the rule of the decrease of Gravity. . . . He says you had the notion from him and seems to expect you should make some mention of him in the preface.' This was too much for Newton to bear. The arrogance and effrontery of the man were astounding.

He wrote back to Halley a letter remarkable, in the period of gentlemanly circumlocution, for its acerbity. He detailed all of his dealings with Hooke, point by point, and rebutted Hooke's claims. 'Now is this not very fine?' he asked Halley. A man who 'does nothing but pretend & grasp at all things' must take the credit for Newton's hard mathematical work. He added that 'he should rather excuse himself by reason of his inability'. Newton was in such a fury that he declared that he would withdraw the proposed third book, telling Halley that 'philosophy is such an impertinently litigious Lady that a man had as good be involved in Law Suits as have to do with her'. This looks very much like one of Newton's fits of pique but, like all such fits, it passed. He was calmed by Halley who seems to have understood the insecure and prickly character of Newton. He begged him 'not to let your resentments run so high'. Within a few months Newton did in fact dispatch the third book to Halley, although he had in the interim decided to couch it in a more difficult mathematical style. Perhaps he hoped that Hooke would not be able to understand it.

Newton was so incensed by the charge of plagiarism, however, that he removed the references to Hooke in his manuscript. One allusion to '*clarissimus Hookius*', the 'most celebrated Hooke', was scratched through. Quite simply, Newton wanted to blot him out. He never forgave him, and he remained the enemy of Hooke for the rest of Hooke's life. In public he advanced to the higher ground of rational debate. Hooke may have guessed or hypothesised various things, he said, but only I have been able to prove them. In the second edition of the *Principia* he declared grandly that '*Hypotheses non fingo*' – 'I do not construct, or feign, hypotheses'. He relied upon mathematical demonstration and proof. He depended entirely upon the 'phenomena' and would not speculate on causes or possible explanations.

So there was one question that remained unanswered. As Newton himself put it in an unpublished preface to the *Principia*, 'I have not yet learned the cause of gravity from the phenomena.' What is gravity? From where does it come? In simple terms, what is it 'made of'? No one is any closer to answering that question. Newton confessed in a later letter that 'the cause of gravity is what I do not pretend [claim] to know', and in a further explanation of his unusual ignorance he declared that 'gravity must be caused by an agent acting constantly according to certain laws, but whether this agent be material or immaterial is a question I have left to the consideration of my readers'.

If Newton did not know the answer, of course, it is unlikely that his readers would be any the wiser. It was enough for him that he had explained its laws. He had explained the mathematics, rather than the physics, of the phenomenon. Of course Newton speculated on the matter – at one time favouring the presence of some 'aether' – but in the end he seems to have believed that, as one associate put it, 'Gravity had its foundation only in the arbitrary will of God'. It is important to recognise that, for Newton, the Divine Being was the begetter and sustainer of the universe; it could not continue to exist without divine intervention. He did not believe in a materialistic, or mechanical, cosmos.

Although the problem of the origin of gravity was not solved, nobody could question the astounding accuracy of Newton's calculations. One French natural philosopher, Fontenelle, declared in an account of Newton that 'Sometimes his conclusions even foretell events which the astronomers themselves had not remarked'. And so it proved. After the publication of *Principia Mathematica*, certain phenomena were observed that validated its conclusions – among them a new measure of the shape of the earth, a more precise observation of the path of Saturn, an analysis of the tides and, most spectacularly of all, the return of Halley's comet in 1758. Newton was the prophet as well as the magus. And his influence has not diminished in the intervening centuries. The American scientists of NASA still depend upon his calculations when launching the latest technology into space.

By the spring of 1687 Newton had despatched the entire

manuscript to Halley. On 5 April Halley acknowledged his receipt of 'your divine treatise' and assured Newton that the 'world will pride it self to have a subject capable of penetrating so far into the abstrusest secrets of Nature, and exalting humane Reason to so sublime a pitch by this utmost effort of the mind'. He addressed his letter to Lincolnshire. Newton, having completed the labour of *Principia Mathematica*, had decided to go home for a period of rest. At a later date he summed up his work in simple terms. 'If I have done the Publick any service,' he wrote, ''tis due to nothing but industry & patient thought.'

Within four months, after much labour and anxiety, Halley had successfully supervised and paid for the printing and publication of the precious manuscript. At the beginning of July a quarto volume appeared from the presses. It was of 511 pages and, bound in leather, could be purchased for nine shillings. The print run was between three and four hundred copies, a small amount for so monumental a work. Halley sent twenty copies down to Cambridge by means of a carrier, so that Newton might distribute them to his colleagues. Their responses are not known. One Cambridge undergraduate, seeing Newton in the street, remarked, 'There goes the man that writt a book that neither he nor anyone else understands.'

That was not exactly true. There were three editions of *Principia Mathematica* – published respectively in 1687, 1713 and 1726 – the later ones benefiting from Newton's revisions and additions. But the effect of his book was, if not widespread, immediate and profound. When the queen of Prussia asked Leibniz for his opinion on Newton's achievement he replied that 'taking Mathematicks from the beginning of the world to the time of Sir Isaac, what he had done was much the better half'. One reviewer of the second edition declared that Newton's computations concerning the motion of the moon 'proves the divine force of intellect and outstanding sagacity of the discoverer'. A mathematician in Scotland wrote to Newton, on receiving the book, thanking him 'for having been at pains to teach the world that which I never expected any man should have knowne' and assuring him that he would receive the plaudits of 'this and all succeeding ages'. John Locke, himself one of the great figures of the age, now considered Newton to be the intellectual master of his generation.

There were some who considered him to be scarcely human, and a French mathematician cried out, 'Does he eat and drink and sleep? Is he like other men?' After the publication of the book Newton recovered his self-confidence and, after his cloistered existence in the years before the *Principia*, positively blossomed in a flurry of correspondence with other philosophers and assorted disciples. The value of his calculations became evident in the fact that 'Newtonianism' had become by the eighteenth century the English orthodoxy.

## Chapter Ten
# The public world

Newton had already become a public figure in another sense. For the first time in his life he had become involved, albeit reluctantly, with national affairs. In the early months of 1687 the new king James II had decreed that the University of Cambridge should admit a Benedictine monk named Alban Francis; he was to be granted the degree of MA without taking the oath of supremacy. This oath, affirming the status of the sovereign as supreme governor of spiritual matters in England, had been in use since its introduction by Elizabeth I in the spring of 1559. But James II, suspected of papist sympathies, seemed willing to forgo the privilege of leading the English Church. The new king had already signalled his willingness to introduce notable Catholics to the universities, and had in fact installed one of them as Master of Sidney Sussex College. Newton made his position clear in a letter to a colleague when he suggested that 'if his Majesty be advised to require a Matter which cannot be done by Law, no Man can suffer for neglect of it'. The king cannot require any subject to break the laws of statute. The 'matter' was of course much more sensitive, since it required the entrance into the university of those whom Newton considered to be bigoted and benighted papists and offspring of the Whore of Rome.

The generally cohesive Protestantism of the university was now put to the sternest test. Could the authorities defy their sovereign and their master in such a matter? If they relinquished the fight without a struggle, then they would be faced with a Catholic presence on a much larger scale. It was at this stage that Newton entered the arena. On 11 March, while waiting for the third book of his *Principia* to be transcribed by his amanuensis and sent to London, he attended a meeting of the university authorities. The terms of his intervention are not known but they must have been powerful enough to mark him out; he was chosen as one of the representatives charged to deliver the opinion that it would be illegal to admit the Benedictine monk.

The king was not at all happy with this suborning of his authority, and in April summoned the leaders of the university before the Court of Ecclesiastical Commission. The vice-chancellor of the university, and the other authorities, were thrown into panic. They feared that their livelihoods might be lost, and their sinecures taken away. At the last minute they prepared a compromise, agreeing to admit Father Francis on condition that their action did not establish any precedent. It was of course a feeble response, and Newton was one of those who spoke against it. His own radical Protestantism was of course not in doubt, and his sturdy dissenting conscience would have given eloquence and authority to any religious opinion he conveyed. He explained to Conduitt at a later date that he had gone up to the beadle, or senior law official of the university, and had said to him, 'This is giving up the question.' Whereupon the beadle replied, 'Why do not you go and speak to it?' So he went back to the table, and argued successfully against the compromise.

So he was one of those despatched to face the fury of the king at the Court of Ecclesiastical Commission, which was at that time guided by the eminent and infamous Judge Jeffreys, otherwise known as the hanging judge, who, according to Gilbert Burnet's *History of My Own Times*, was 'either perpetually drunk or in a rage'. Newton and his eight colleagues appeared four times before him, when he seems to have been in the latter condition. He accused the vice-chancellor of the university, John Peachell, of 'an act of great disobedience' and promptly stripped him of his position and of his income. It seemed likely that a similar punishment would await Newton and the others, but Newton's resolve seems to have been strengthened. He added a paragraph to a declaration by the university to the effect that 'A mixture of Papist & Protestants in the same University can neither subsist happily nor long together. And if the fountains once be dryed up the streams hitherto diffused thence throughout the Nation must soon fall of.' Fortunately for him, perhaps, the declaration was never delivered.

At their final meeting with Jeffreys the judge decided to deal mildly with them, and blamed their defiance upon their erstwhile vice-chancellor. 'Therefore I shall say to you what the scripture says, and rather more because most of you are divines; Go your way, and sin no more, lest a worst thing come unto you.' So they journeyed back to Cambridge relatively unscathed; more significantly, the case of Father Francis was not further pressed by the Crown. James must already have realised that the climate of opinion in the country was turning against him. Less than two years later, Jeffreys himself died in a prison cell.

Yet despite the perilous circumstances Newton's first experience of the public world seems to have sharpened his appetite for life outside the confines of the university. The arrival of Mary II and William of Orange, the stadtholder of the Netherlands, and the renewal of the Protestant ethic at Whitehall and elsewhere, presented him with an opportunity. In fact the 'glorious revolution' of 1688, during which James II was deposed before Mary and William assumed the throne in sternly Protestant guise, came as a marked relief to Newton himself. On 15 January 1689 he was elected as one of the two representatives of the university to take part in the national convention that would ratify the quiet 'revolution' in national affairs. His firm and articulate interventions in the earlier controversy over the Benedictine monk had been remembered. He was also now benefiting from the glory and prestige that accrued to the author of the newly published *Principia Mathematica*.

On 17 January, two days after his election, he was dining with William of Orange in London. It was a sudden emergence in London life, but he seemed to enjoy his new eminence. He remained in London for a further twelve months, with a brief respite when the convention went into adjournment. The convention itself became a parliament after the Crown had been formally granted to William and Mary, in the middle of February. So Isaac Newton had become an MP. He did not take much part in the deliberations, however, and it is said that he only spoke out once – when he asked an usher to close a window for fear of a draught. He may have had some reason to do so. He suffered from some undefined indisposition in March, and two months later contracted a 'cold & bastard Pleurisy'. The

Newton was born in the manor of Woolsthorpe. The orchard was the setting for the legend of the apple which fell from the tree, while Newton was at Woolsthorpe escaping from the plague of 1665; he himself recounted four different versions of this story to four separate people.

Dr Isaac Barrow, first Lucasian Professor of Mathematics at Cambridge. He was Newton's tutor, and allegedly 'reckon'd himself but a child in comparison of his pupil'.

(*Above left*) Edmond Halley. Knowing that Newton lived a cloistered existence, he simply took a coach to Cambridge and bearded him in his den, where he asked him a momentous question about the movements of the planets. Newton claimed already to have calculated the answer.

(*Above right*) The Astronomer Royal, John Flamsteed. Newton treated him like a technical assistant. 'I want not your calculations,' he wrote, 'but your Observations only.'

Plate from Flamsteed's 'Atlas Coelestis'. Newton was highly obstructive about its publication; he wanted Flamsteed's research for his own star maps. 'Sr I N having been furnished from My Stores,' wrote Flamsteed, 'would charitably bestow it on the publick that he may have the prayse of having procured it.'

Nicolas Fatio de Duillier.
Newton seems to have had an affection
for this bumptious young man whom he
met at a meeting of the Royal Society.
Fatio, in turn, reported himself 'frozen
stiff' by Newton's genius.

Sir Isaac Newton.
Even as a young man, he was
numbered among his peers as one of
the 'Worthies of Britain'.

Robert Boyle. Newton made
many notes on Boyle's *Experiments
& Considerations touching Colours*,
followed by his own experiments
on the nature of light. He stared so
long at the sun with one eye that
he had to spend three days in a
darkened room to restore his sight.

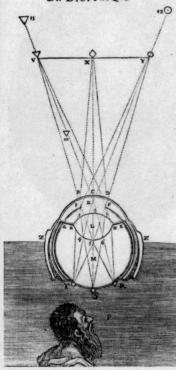

Image from Descartes' *Dioptrique*.
To test Descartes' theory that light was
a 'pressure' pulsating through the ether,
Newton inserted a bodkin 'betwixt my
eye and the bone as near to the backside
of my eye as I could'.

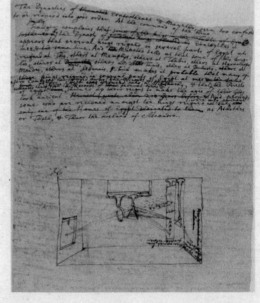

Newton's drawing of his
sunlight refraction experiment.

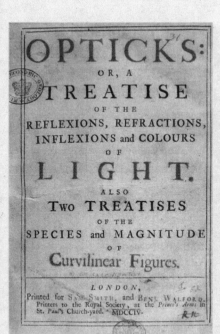

**OPTICKS:**
OR, A
**TREATISE**
OF THE
REFLEXIONS, REFRACTIONS,
INFLEXIONS and COLOURS
OF
**LIGHT.**
ALSO
Two **TREATISES**
OF THE
SPECIES and MAGNITUDE
OF
**Curvilinear Figures.**

LONDON,
Printed for Sam. Smith, and Benj. Walford.
Printers to the Royal Society, at the Prince's Arms in
St. Paul's Church-yard. MDCCIV.

Newton's *Opticks* 'set forth
unheard of wonders about colours'.

Portrait of Newton on a half-penny.

Newton's statue at his college,
Trinity, Cambridge.

James Thornhill's portrait of Newton.

Lord Halifax, Chancellor of the Exchequer, who gave Newton his position as Warden of the Mint.

The Tower of London and the Mint, during Newton's tenure as Warden.

Iron cast of Newton's death mask.

Newton's friend, William Stukeley.
He collected all the material he
could find on his hero and wrote
a memoir of him.

atmosphere of London was not a healthy one.

He rented lodgings in Broad Street, Westminster, close to the House of Commons, for at least some of the time. He seems to have entered the spirit of the busy city, and steadily enlarged his circle of acquaintance. Among his new companions was John Locke, the philosopher, as well as assorted Whig notables. Locke himself endeavoured to bring the orderly world of science into his discussions of social philosophy, and in Newton found a fellow enthusiast of an empirical and experimental temper. He even went so far as to describe his new friend as 'the incomparable Mr Newton'.

Newton attended meetings of the Royal Society, despite the presence of Hooke, and that summer he met there the Dutch natural philosopher, Christiaan Huygens, whose own work on light and gravity made him the only man in Europe to come close to Newton's achievement. At the Royal Society, too, he made the acquaintance of Samuel Pepys. Newton also confirmed his friendship with Charles Montagu, who had been a Fellow at Trinity, but who was now in the process of climbing up the ladder of political preferment. He would turn out to be a key influence on Newton's own future.

His new confidence is marked by the fact that he was painted by Sir Godfrey Kneller, the foremost portraitist of the period. It is perhaps surprising that Newton should acquiesce so willingly in being painted, and the Kneller picture is only the first of a number of portraits of the great man. But his vanity seems to have surmounted his scruples in this matter. He was aware of his achievement, and was happy to be commemorated for it.

It is the portrait of a man who knows his worth. He has long hair, grey and silver, and his glance is at once keen and inclusive. His eyes are slightly protuberant, the product of myopia and long experimental observation. He seems to have been caught in a moment of thought, his full mouth and prominent nose adding to the impression of contemplation not untouched by anxiety. He is wearing a linen shirt and a loose academic gown. His presence in the world is decisive, almost commanding. It was the first of three portraits of Newton that Kneller executed, and by the second of these paintings it is possible to see Newton become visibly more magisterial and domineering. There are records of some seventeen portraits of Newton altogether, by any standard a great number for even the most famous natural philosopher. Yet he wished for them. Perhaps they helped to validate his identity; perhaps they gave him visible status in a world of signs and symbols. Perhaps he had a strain of narcissism within his temperament. A person who is not known to have had any very strong emotional or sexual attachment to any other human being might, conceivably, fall in love with himself.

## Chapter Eleven
# Hero worship

He did form an attachment of some kind, however, to a young man whom he met during this first year in London. Nicolas Fatio de Duillier was an engaging and enthusiastic young man of Swiss descent; he was twenty-five when he met Newton at a meeting of the Royal Society, but he was already adept in mathematics and astronomy. He seems to have ingratiated himself with the older man, and Newton was suitably charmed by his cleverness and ready wit. He was so impressed by him, in fact, that in the autumn of the year he wrote that 'I . . . should be very glad to be in the same lodging with you. I will bring my books & your letters with me.' He also wrote to Fatio in criticism of Robert Boyle, whom he accused of 'being in my opinion too open & too desirous of fame'. It was quite unlike Newton to be so unguarded in his correspondence, and suggests a considerable amount of trust and affection on his part towards this young Swiss mathematician.

Fatio in turn seems to have considered his new friend to be a hero and semi-divinity, with whom it was a privilege to discourse. He also set about offering him some practical assistance, and employed his friendship with John Locke to advance his cause. 'I did see Mr Lock,' he wrote to Newton, '. . . and I desired earnestly that he should speak earnestly of you to Mylord Monmouth' concerning some political appointment. The busy young man had also formed a friendship with Huygens, and offered to send Newton a copy of the physicist's newly published *Traité de la Lumière*. 'It being writ in French,' he added, 'you may perhaps choose rather to read it here with me.'

Fatio had his own theories on the physical nature of gravity, and asserted without much proof that Newton had agreed with his findings. This is unlikely to have been the case, and indeed another mathematician recorded later that 'Mr Newton and Mr Hally laugh at Mr Fatio's manner of explaining gravity'. The young man also claimed that he knew 'no one who so well and thoroughly understands a good part [of the *Principia*] as I do', and seriously considered the possibility of adding material to that volume in a style more easy to understand. His vanity and bravura may have been considered charming. Certainly Newton found the young man endearing enough to maintain the friendship and, after Fatio's prolonged absence in Europe, he wrote to John Locke in the autumn of 1690 asking for news of him. Fatio did in fact return to England in the following year, and seems to have met Newton often both in London and in Cambridge. When Newton showed him some of his mathematical calculations Fatio reported that he was 'frozen stiff' at the genius showed by them.

In the letter to Locke, in which Newton had inquired after Fatio, he also raised the question of political preferment in London. He was interested in the position of comptroller of the Mint, the institution that controlled the minting of coin for the whole of the country, but for the time being he was not successful. He was later recommended by Locke for the vacant position of Master of Charterhouse alms-house and school, accommodating eighty poor men and forty foundation scholars, but he did not think the place worthy of him; he had been told that 'its but 200£ *per an* beside a Coach (which I reccon not) & lodgings'. He had an eye upon grander, and better rewarded, employment. He added that 'the confinement to the London air & a formal way of life is what I am not fond of', but this caveat need not be taken seriously. After his introduction to the political and public world he seemed anxious to return to it. The air of London may not have been to his liking, but the atmosphere of power and profit certainly was. He may have felt constricted, and neglected, in his Cambridge rooms. It is perhaps odd that he did not feel that he had accomplished enough with his theory of universal gravity and with the knowledge that he had become the most eminent scientist (or natural philosopher) in the world. He wanted more.

He may have realised, also, that his days of original thought and laborious calculation had come to an end. He was now in his late forties, past the prime of mathematical genius, and saw nothing ahead of him but a lonely and morose existence in a Cambridge college. He remained active in his alchemical researches and in his investigations of Scripture, but it is clear that he wanted to get out, while there was still a chance to do so. He wanted to succeed in quite another way of life. And, in the end, he proved that he could.

In the spring of 1692 a young divine, Richard Bentley, gave a series of sermons at St Martin-in-the-Fields on what he called 'A Confutation of Atheism'. In the course of these sermons he used Newton's recent discoveries, as outlined in *Principia Mathematica*, as a way of confirming the workings of divine providence in the universe. Before committing his work to the press for eventual publication, he took the precaution of writing to Newton himself for confirmation of certain matters. Newton wrote back with the firm declaration that 'when I wrote my treatise about our Systeme I had an eye upon such Principles as might work with considering men for the beliefe of a Deity & nothing can rejoice me more than to find it usefull for that purpose'.

Are we then to consider *Principia Mathematica* as a religious, as well as a scientific, treatise? That may be going too far, given the rigorous and even forbidding mathematical content, but there can be no doubt that for Newton natural philosophy had an innately religious purpose. In his letters to Bentley he declared that at the moment of creation God formed every particle of matter 'with an innate gravity towards the rest'. The theory would imply that all matter would eventually coalesce, except for the fact that the divine being had made the universe infinite. So some parts of matter 'would convert into one mass & some into another so as to make an infinite number of great masses scattered at great distances from one to another throughout all that infinite space'. This is in part the modern vision of the universe.

Once again he made it clear to Bentley that gravity was not inherent in matter but, rather, had been created by some unknown agent. He stated that 'the cause of gravity is what I do not pretend to know', except that it required the 'mediation of something else which is not material'. He added that this unknown cause was also 'very well skilled in

Mechanicks and Geometry'. That was not blasphemous, only a reminder that mechanics and geometry had a divine foundation.

It was often supposed, by William Blake and the Romantic poets for example, that Newton was the one man who stripped the universe of divine causation and who rendered it wholly mathematical. Nothing could be further from the truth. He insisted that the universe could only be understood as the work of a Creator, and that its orderliness was the result of a divine plan.

He had other acolytes of a more empirical temper than Richard Bentley, who assumed the role of disciples and spent their professional careers in the dissemination of Newtonian theory. One of the most eminent was David Gregory, a Scottish mathematician who as Professor of Mathematics had promulgated his work at the University of Edinburgh. In 1691, partly as a result of Newton's recommendation, Gregory was appointed Savilian Professor of Astronomy at Oxford. Newton was intent upon placing supporters of his theories in the most prominent and influential posts. It was part of his need to manipulate and to control the world.

Another Newtonian enthusiast was William Whiston; he had heard Newton's lectures as an undergraduate and his passion for Newtonian mathematics was only equalled by his espousal of Newton's rigorous Arianism. He eventually became Newton's successor as Lucasian Professor of Mathematics, having been appointed by Newton, although his tenure of this office turned out to be an unhappy one. Later disciples included Colin Maclaurin and Henry Pemberton; one became Professor of Mathematics at

Edinburgh, the post that Gregory had once filled, and the other was appointed Professor of Physic at Gresham College. Newton of course was instrumental in their advancement. He also arranged for Edmond Halley, one of his principal admirers and exponents, to become Savilian Professor of Geometry at Oxford. After his death, his scientific orthodoxy was set in place.

One of his disciples, however, seemed to be in peril. On 17 November 1692 Fatio de Duillier pleaded sickness unto death. 'I have Sir,' he wrote, 'allmost no hopes of seeing you again.' He explained that, on leaving Cambridge after his last visit to Newton, he had caught a cold that speedily fell upon his lungs and fostered an 'ulcer'. 'I thank God my soul is extreamly quiet, in which you have had the chief hand. My head is something out of order, and I suspect will grow worse and worse.' He complained that the medicine he had taken, the 'Imperial powders', 'have proved quite unsignificant'. He added then, in a somewhat rhetorical flourish, 'were I in a lesser feaver I should tell You sir many things. If I am to depart this life I wish my eldest brother, a man of extraordinary integrity, could succeed me in your friendship.' Then, paradoxically, he concluded 'As yet I have no Doctor'.

The odd concept of his friendship with Newton being a hereditary position is matched by his evident lack of interest in summoning specialist help. The missive has often been dismissed as the product of an hysterical and hypochondriacal imagination, looking for sympathy and affection by the most effective means possible. But that is perhaps to understate the prevalence of fatal disease in the period, as well as the evident hopelessness of recourse to doctors.

There were many people who, feeling seriously unwell, prepared for the worst.

Certainly Newton's response was fearful and immediate. He 'last night received your letter with which how much I was affected I cannot express. Pray procure the advice & Assistance of Physitians before it bee too late & if you want any money I will supply you.' He added that he would form the acquaintance of his brother, although 'I hope that you may still live to bring it about, but for fear of the worst let me know how I may send a letter &, if need be, a parcel to him'. He signed himself 'your most affectionate and faithfull friend'. It was the most 'affectionate' letter he ever delivered, and testifies to the depth of his concern for the young man, but it does not shrink from envisaging 'the worst'. That is a measure of Newton's piety and hope for a better place after the world's adventures.

Fortunately his anxieties were misplaced. Five days after sending his first distraught message Fatio wrote to inform Newton that 'I hope the worst of my disease is over. My lungs are much better . . .' This hopeful note was followed by a long record of his symptoms, as if to convince Newton of the inherent seriousness of his case. In the following January, Fatio's illness not quite dissipated, Newton invited the young man to stay with him in Cambridge. 'I feare the London air conduces to your indisposition,' he wrote, '& therefore wish you would remove hither so soon as the weather will give you leave to take a journey.'

Fatio then replied with the no doubt unwelcome news that he would have to depart for his native Switzerland after the recent death of his mother; he had been left an inheritance which needed his attention. But he assured Newton that, if he had been left enough money, he would prefer to live in England and 'chiefly at Cambridge, and if You wish I should go there and have for that for some other reasons than what barely relateth to my health and to the saving of charges I am ready to do so; But I could wish in that case You would be plain in your next letter'.

It has been surmised or hinted that there is some sexual innuendo in this request concerning 'some other reasons'. But that is unlikely to be the case. Fatio was concerned both with Newton's theological and mathematical researches, and was here adverting to the prospect of shared work. In the same letter he refers to the biblical prophecies, which he believed to relate 'most of them to our own times and to times lately past or to come'. In future years Fatio would indeed become a religious enthusiast of the most extreme kind. Even at this early date Newton sensed some lack of balance in the young man, and in a reply to his letter warned him that 'I fear you indulge too much in fansy in some things'.

Their correspondence continued over the next few months. Newton sent Fatio money and promised, if the young man decided to reside in Cambridge, 'to make you such an allowance as might make your subsistence here easy to you'. Fatio replied in fulsome terms. 'I could wish Sir, to live all my life, or the greatest part of it, with you.' So Newton was capable of inspiring, as well as demonstrating, affection.

The last extant letters from Fatio are, intriguingly, concerned with alchemical experiment in which he seems to have considered himself an adept. They are intriguing because they emphasise the fact that Newton himself was still deeply involved in occult studies and discussed them in detail with his young disciple. Fatio's message at the end of one letter was 'to burn this letter after you have done with it'. In the last of all the letters Fatio revealed that he had come upon a secret elixir that would promote physical health. 'I could cure for nothing thousands of people and so make it known in a little while. After which it would be easie to raise a fortune by it.' Newton may not have been offended by Fatio's plea for financial investment in the scheme, but he must have realised the over-credulous nature of his friend's enthusiasms.

The elixir might have been intended for Newton himself. From the autumn of 1692 he felt himself to be suffering from increasing ill-health. He was anxious and ill at ease. He suffered from insomnia. In the summer of 1693 he wrote a short alchemical treatise that was eventually entitled *Praxis*; it is clothed in alchemical language, as, for example, in the description of two 'immature substances' that 'become a pure milky virgin-like Nature drawn from the menstruum of the sordid whore'. Other substances 'will then become oyles shining in the dark & fit for magicall uses'. This is the language of the magician as much as of the experimenter, if the titles can readily be distinguished. On two occasions in May and June he travelled to London for short visits, perhaps to see Fatio, but his poor health was not alleviated. Then, as if struck by sudden lightning, his mind was plunged into chaos.

## Chapter Twelve
# The balance is lost

The first example of Newton's mental collapse emerges in a letter that he sent to Samuel Pepys on 13 September 1693, concerning his attempts to find public preferment. 'I am extremely troubled by the embroilment I am in,' he wrote, 'and have neither ate nor slept well this twelve month, nor have my former consistency of mind.' He went on to declare that 'I never designed to get anything by your interest, nor by King James's favour, but am now sensible that I must withdraw from your acquaintance, and see neither you nor the rest of my friends any more . . .'. He signed himself 'your obedient servant'.

In the next day or two he travelled to London and, while lodging at the Bull Inn in Shoreditch, sent a similarly unbalanced letter to John Locke.

> Sir, being of opinion that you endeavoured to embroil me with woemen [an odd if for him appropriate spelling] & by other means I was so much affected with it as that when one told me you were sickly & would not live I answered twere better if you were dead. I desire you to forgive me this uncharitableness.

In the midst of what can only be described as rant he added

that 'I beg your pardon also for saying or thinking that there was a designe to sell me an office, or to embroile me'. It is sometimes wondered what he was doing in Shoreditch, so far from his previous haunts in the West End and Whitehall. The Bull Inn, however, was the regular port of call for travellers from the eastern counties. Yet it is still somewhat odd to take the carriage from Cambridge and then, at the coaching inn, to write the letter.

Both letters were in fact exceedingly odd, and the recipients were alarmed at Newton's tone. Pepys made discreet enquiries at Cambridge and a university Fellow, who was a friend of Pepys's nephew, was prevailed upon to visit Newton. He reported that 'before I had time to ask him any question, he told me that he had writ you a very odd letter, at which he was much concerned'. He had 'added that it was in a distemper that much seized his head, and that kept him awake above five nights together'. He begged Pepys's pardon and the informant went on to note that 'he is now very well and, though I fear he is under some small degree of melancholy, yet I think there is no reason to suspect that it hath at all touched his understanding'. So he was subdued, and perhaps depressed, but there were no signs of mania. Pepys replied to Newton with an affable letter, but he also did a shrewd thing. He put to Newton a question of calculation, upon the chances of throwing sixes at dice in various combinations. This may have been a simple enquiry. What good was a mathematical genius if he could not give advice about the gaming table? But it seems more likely that Pepys was subtly testing Newton's mental powers in the wake of the strange outburst. Newton duly answered, with reassuringly accurate calculations.

John Locke replied to Newton after a fortnight in a dignified letter, expressing 'hopes I have not lost a freind I so much valued'. In his reply Newton confessed that 'last winter by sleeping too often by my fire I got an ill habit of sleeping & a distemper which this summer has been epidemical'. He added that 'when I wrote to you I had not slept an hour a night for a fortnight together & for 5 nights together not a wink . . . I remember I wrote to you but what I said of your book I remember not'. He had in fact said to Locke that in his book he 'struck at the roots of morality'. Morality had been much on his mind in this period of sleeplessness. In both of the original letters to Locke and Pepys he reiterated his fear of 'embroilment', in particular with 'woemen', which suggests a deep suspicion of sexuality. It was the enemy.

There have been many accounts of Newton's 'breakdown', as it has been called, some of them contemporaneous with the affairs in question. One Scottish correspondent told Huygens that Newton, after a fire among his papers and experiments, had suffered an attack of frenzy that had lasted for eighteen months. The news spread through Europe, and two years after the event a German philosopher was writing that a fire in Newton's lodgings made him 'so disturbed thereupon, as to be reduced to very ill circumstances'.

There had been a fire, or perhaps fires, in the past. No alchemist could be entirely free of them. But they seem to have played no part in Newton's illness. The reports of his 'breakdown' were also much exaggerated. Far from being incapacitated for eighteen months, he seems according to Pepys's informant to have recovered much of his equilibrium some two weeks after the original onset of the problem – whatever that problem was.

There have been many explanations. It has been suggested that a lifetime of hard work and thought had cost him temporary loss of his sanity; this is certainly possible. Intense and laboured meditation can provoke severe depression. No man had worked harder, or more rigorously, over the problems of the universe. Newton may also have suffered nervous anxiety from the publication of *Principia Mathematica*, when his work was as it were revealed to the world; his anxiety may also have been deepened and strengthened as a result of his test of wills with the previous monarch. His whole livelihood had then been under threat. In more recent years he had also been frustrated in his search for preferment, a central point in his troubled letters, and the anxiety and depression attendant upon that pursuit may have touched him.

It has also been suggested that a lifetime of alchemical experiment may have affected his mental health, particularly because of his prolonged exposure to mercury. Some of the symptoms of mercury poison are sleeplessness and a tendency to paranoid delusions, both of which apply in Newton's case. Modern researchers have in fact found large quantities of lead and mercury in the remaining traces of his hair, kept over the centuries by family members. But he suffered no other obvious symptoms, such as tremulousness and rotting teeth. There is also the evident fact that Newton recovered quickly, whereas poisoning by mercury is a long-term and insidious process. He is also likely to have known the effects of such poisoning, and would not have described them as an epidemical distemper.

There is of course the imponderable effect of his friendship with Fatio. The emotional temperature of their relationship was high, higher than usual among adult males of the period, and it may be that Newton's panic fear of 'embroilment' applied to the young man also. It is pertinent that in his odd letter to Locke he had written of his attempt 'to embroil me with woemen & *by other means*' (my italics) – Locke and Fatio had been staying together with a mutual acquaintance, Lady Masham, when they invited Newton to join them. It is certainly true that, in the course of this unhappy period, Newton and Fatio broke off their friendship of four years; it was never renewed. The rest must be merely speculation.

Whatever the cause of his temporary lapse, it is clear that Newton recovered from it quickly. His mental and physical resilience was in fact remarkable, a sure testimony to the reserves of energy and health that he brought to his intellectual work. In the autumn of 1693 he resumed his correspondence, apologising to Huygens among others that he had mislaid their original letters. In the course of his letter to Huygens he claimed, somewhat implausibly, that 'I value friends more highly than mathematical discoveries'. He may, however, have been chastened by his recent experience.

In succeeding months he was conversing with David Gregory (the newly established Professor of Astronomy at the University of Oxford) on matters mathematical and cosmological, and planning with him alterations for a new edition of *Principia* which would entail an entire restructuring of its contents; he also wanted to add a treatise on the geometry of the ancients, to confirm his own belief that the secrets of the universe were already known to his distant progenitors. He was also considering a volume concerning his optical discoveries, which eventually emerged in the world as *Opticks*. Gregory noted that such a book 'if it were printed would rival the *Principia Mathematica*. . . . He sets forth unheard-of wonders about colours.'

In the same period he was collating manuscripts for his study of the Apocalpyse, and drawing up a mathematical prospectus for Christ's Hospital school. He was still engaged in alchemical readership, part of the obsession that never left him. This was not a man in the wake of an extended or extensive nervous breakdown.

Certainly he had not become more credulous. There had been reports of a ghost inhabiting a house opposite the gate of St John's, just down the street from Newton's own rooms in Trinity. Several scholars and fellows of the university visited the dwelling in hope of strange phenomena. Newton saw them, as he walked past, and according to report felt moved to expostulate: 'Oh! Yee fools, will you never have any witt, know you not that all such things are meer cheats and impostures? Fy, fy, go home, for shame.'

The best evidence that Newton had lost none of his rational capacity lies in the fact that in 1694 he was also renewing his work upon what he called 'the theory of the Moon'. The moon remained one of his obsessions, and he once confessed that 'his head never ached but with his studies on the moon'. The problem of its motion, in relation to his theory of gravitation, was one that he would never be able to solve. Yet he was nothing if not determined; in theory at least, no problem was incapable of solution. With David Gregory he visited the Greenwich Observatory at the beginning of September 1694, in pursuit of more accurate lunar observations.

The Astronomer Royal, John Flamsteed, was a cautious and defensive man. He was largely self-taught, and had surmounted a childhood of poverty and ill-health to be appointed by Charles II as the first Astronomer Royal. The king had not willed him the means, however, to fulfil his office properly. He had little money, and the observatory was chronically short of resources and equipment. Nevertheless he persevered. His great project was a 'star catalogue' that would fix the positions of all the known stars. But he also had a strong desire to ingratiate himself with Newton; he wrote in a memorandum that his 'approbation is worth more to me than the cry of all the Ignorant in the World'. Accordingly he showed to Newton the results of his lunar observations, on condition that he revealed them to no one else.

Newton was a hard and imperious colleague, however, who continually asked Flamsteed for more observations and questioned those that did not correspond with his theories. When Flamsteed asked for some elucidation of these theories, he proved incapable of understanding them. He also made the mistake of questioning Newton's accuracy in certain calculations. This made Newton even more impatient, and he asked Flamsteed simply to send the raw data without further comment. 'I want not your calculations,' he wrote, 'but your Observations only.' He had also offered to pay Flamsteed for his work, which Flamsteed deemed undignified and insulting: 'I am displeased with you not a little,' he wrote, 'for offering to gratifie me for my paines.' It seemed that Newton was treating him as a menial, or mere technical assistant, in offering him payment.

Flamsteed believed himself to be contributing to the work of an intellectual peer.

Newton had assumed that he could finish his theory of the moon in relatively short time. 'I reccon,' he had written to Flamsteed, 'it will prove a work of about three or four months & when I have done it once I would have done with it forever.' There is the essential tone of the man, busy and expeditious. Once he had completed a task, he dismissed it from his mind.

But Flamsteed was not providing the materials that Newton needed to finish his work satisfactorily. When Flamsteed complained of illness Newton was not particularly sympathetic; on learning of the astronomer's headaches he advised him to bind his head with a garter until it was 'nummed'. In an angry and impatient letter Newton wrote to Flamsteed that 'when I . . . saw no prospect of obteining them or of getting your Synopses rectified, I despaired of compassing the Moon's Theory, & had thoughts of giving it over as a thing impracticable'. Then he stopped answering Flamsteed's letters. He erupted into a further fit of fury on the discovery that Flamsteed had divulged some of his lunar calculations, and expressed his abhorrence of being 'dunned & teezed by forreigners about Mathematical things'.

He did eventually print some further calculations concerning the moon's trajectory, in the second edition of the *Principia*, but he never once mentioned John Flamsteed. He even expunged previous references to the astronomer's name. In a private note Flamsteed described Newton as 'hasty, artificial, unkind, arrogant'. The two men were to engage in an even more bitter controversy ten years later. It was the pattern of Newton's life.

## Chapter Thirteen
# Matters of coinage

Newton had in part lost interest in the moon because of his concern with more worldly matters. His friend and erstwhile Cambridge contemporary, Charles Montagu, had been appointed Chancellor of the Exchequer and thus possessed considerable powers of patronage. Why should Newton not benefit? It seemed just, and natural, that the foremost mathematician of the age should play some part in administering the nation's economy. There had already been rumours and reports about his possible employment at the Mint in London, rumours that he had dismissed out of hand. But then in the spring of 1696 Montagu offered Newton the post of Warden of the Mint. He accepted at once, and within a month had departed from Cambridge to London without any sentimental leave-taking. He must have left in a hurry because, after his death, his college rooms were shown to visitors complete *as he had inhabited them*.

He seems to have become heartily sick of Cambridge and its inmates. Although for the next five years he still held tenure as Fellow of Trinity and Lucasian Professor of Mathematics, he returned to his old university for only three or four days during that period. He had in the past needed the seclusion and isolation of the city in the fens in order to prosecute his work, but the days of concentrated intellectual endeavour were over. He had never sought out the company of his fellow academics, and seems to have found his real intellectual stimulation at the Royal Society in London. Even his fellow alchemists were more easily to be found in the capital – just before his acceptance of the post at the Mint, a mysterious visit from 'A Londoner' was concerned with occult consultations on the 'menstruum' that dissolved all metals. Their conversations lasted two days. So London was the lodestone, the magnet, of Newton's activities and ambitions.

Newton joined the institution at a moment of national economic crisis. In the autumn of the previous year, along with other leading figures, he had written a short treatise entitled 'On The Amendment of English Coyns'. The problem was simple. There were too many handmade silver coins in circulation, clipped and diluted in silver content, and the recent issue of machine-made coin had not appreciably helped the situation. Something like 95 per cent of the currency was counterfeit or under weight. It had been decided that a massive effort of recoinage was necessary, and that the old handmade coins should be removed from use. Newton did not initiate this policy, but he was the person chosen to administer and organise its implementation. It seems that he was offered the post as a sinecure, Montagu having told him that it 'has not too much bus'nesse to require more attendance than you can spare', but it was not in Newton's nature to be anything other than rigorous and determined in all of his activities. In fact at a later date Montagu himself confessed that he could not have managed the change of coinage without the intervention and assistance of Newton.

He remained at the Mint for the rest of his professional life, and his formative intellectual activity came to an end after his removal to London. But his nature, and temperament, did not change. He made it his responsibility – indeed his duty – to familiarise himself with every detail of the Mint's business, to imbibe all the economic theories of the time, and to become acquainted with the history of coinage. He even investigated the various royal warrants issued to the Mint over the previous two centuries. He fashioned every aspect of his new business into order and regularity. That was why he was also a good administrator, understanding intimately every facet of the work. He was an alchemist, for example, and therefore comprehended the intricacies of metallurgy. He was also an exacting master, as his employees soon discovered. He jotted down in one notebook that 'two mills with 4 millers, 12 horses, two horse-keepers, 3 cutters, 2 flatters, 8 sizers, one nealer, three blanchers, two markers, two presses with fourteen labourers to pull at them can coin after the rate of a thousand weight or *3000 lb* of money per diem'. We can be sure that he told the flatters and cutters exactly how much was expected of them.

Such a conscientious and energetic man would inevitably come into conflict with his superior. Thomas Neale was Master of the Mint, a man whom Newton described as 'in debt & of a prodigal temper & by irregular practices insinuated himself into the Office'. Neale was in fact a placeman, an idle and incompetent official who used his public position for the purposes of private enrichment. In this he was no different from most other public servants of the period. But Newton was not of that stamp. He had brought order and certainty out of chaos in the universal sphere, and was not to be impeded in the smaller world of the Mint. He set about increasing his power (and his salary) while at the same time he gradually took command of all the Mint's operations. At the time these employed some five hundred men. It was his nature to dominate and to control.

The recoinage for the entire nation did not of course proceed smoothly. There was a shortage of coin in the first months, before Newton's arrival, and the presses were put in service from four in the morning to midnight. Five months after his arrival, they were manufacturing some £150,000 worth of silver coin each week. Newton's role as warden also included the pursuit and capture of coiners and counterfeiters; he had in effect to become a kind of detective, investigating the nefarious activities of the 'clippers' as they were known. It was unrewarding and thankless work, often impeded by the understandable reluctance of courts to rely on the evidence of paid informers. In one memorandum to the Treasury he complained that 'this vilifying of my Agents & Witnesses is a reflexion upon me which has gravelled me & must in time impair & ruin my credit'. He complained, too, of the calumnies of 'Coyners & Newgate Sollicitors', an apt indication of the company he was now obliged to keep in furtherance of his duties.

He did in fact become a model investigator, as might have been expected of a man who had successfully investigated the cosmos, and pursued his prey assiduously and rigorously. He raided the premises of counterfeiters with his officers, personally interrogated them and visited them in their cells at Newgate and elsewhere. Another official commented that he 'took all informations of which wee burnt boxfuls – & attended all their trials'. His expenses included costs 'in Coach[-h]ire and at Taverns, Prisons and other places in the Prosecution of Clippers and Coyners'. He hired agents in eleven different counties to track down the guilty, and was himself made a Justice of the Peace in the Home Counties to bolster his own efforts.

It is perhaps not surprising that the coiners themselves began to grumble about him with especial loathing for his zeal and tenacity. One such counterfeiter was reported as complaining 'of the Warden of the mint for severity against Coiners and say Damne my blood I had been out [of Newgate] before now but for him'. Another prisoner, caught in Newton's net, declared that 'the Warden of the mint was a Rogue and if ever King James came againe he would shoot him', to which his cell-mate replied, 'God damn my blood so will I and tho I don't know him yet Ile find him out.' One of the most notorious coiners, William Chaloner, stated that 'he would pursue that old Dogg the Warden to the end so long as he lived'.

It was fortunate for Newton that Chaloner did not live very long: three months later he was 'drawn on a sledge' to Tyburn where he mounted the fatal tree. Newton may have taken especial pleasure in his fate because Chaloner had previously informed a parliamentary committee that he knew a much better method of coining money than that employed by Newton, and even offered himself as Supervisor of the Mint. The committee then ordered Newton to study Chaloner's methods of coining, to which Newton took violent objection; it would have meant revealing to the counterfeiter the secrets of the Mint. Eventually Newton prevailed. It would seem, however, that he pursued these coiners at no inconsiderable danger to himself. But who would doubt that he, of all people, would persist in his work?

It is a little disconcerting to consider the author of *Principia Mathematica*, the greatest scientific genius of his age, treading down the stone passages of Newgate Prison or listening to the whispered evidence of those condemned to die; (the penalty for counterfeiting the coinage was death by hanging); it seems to be material more for the novelist than the biographer. Yet the life of an extraordinary human being is bound to manifest extraordinary incongruities. Or is there perhaps some especial congruity at work? The secretive and obsessive philosopher, the alchemical adept, the man who made few friends, does not seem at all dissimilar from the zealous interrogator of those about to hang. We surmise the same intensity, the same rapt concentration upon the task, in both spheres of his life.

## Chapter Fourteen
# Female company

The buildings of the Mint were part of the Tower of London, a necessary precaution against theft or violent riot. Newton took occupation of the warden's house, situated between the outer wall and the bailey of the Tower, in a cramped and noisy position. Five months later he had moved to the more salubrious neighbourhood of Jermyn Street, in Westminster, where a plaque can still be seen commemorating his residence. His life in London was not without its pleasures, despite his onerous responsibilities at the Mint. One of his colleagues reported that 'he always lived in a handsome generous manner tho' without ostentation or vanity, always hospitable & upon proper occasions gave splendid entertainments'. He lived up to his new social position, in other words, and was neither mean nor avaricious. He eventually possessed a private coach, and employed some six household servants. Among his possessions were two silver chamber-pots.

He was elected to the Council of the Royal Society in the year after his arrival in London but played no real part in its proceedings until the death of his enemy Hooke. He could not brook challenge, interference, or even the suspicion of rivalry. His solitary state was in any case lightened by the arrival of his niece, Catherine Barton, to act as housekeeper for his home in Jermyn Street. She was the daughter of his half-sister, Hannah Smith, who had fallen on hard days. It seems likely that Catherine entered the household soon after Newton's purchase of the house in 1696, when she was sixteen years of age. It may seem odd that he should decide to cohabit with a young female relative but, for life in London, a housekeeper was a necessity; in the absence of a wife, what could be better than a young and malleable niece?

Catherine Barton is also reputed to have been beautiful, charming, and of great natural wit. She was toasted by the Kit-Cat Club, with certain verses beginning:

> At Barton's feet the God of Love
> His Arrows and his Quivers lay . . .

She also became an intimate friend of Jonathan Swift who declared that 'I love her better than any one here . . .'. She must have been an extraordinary young woman indeed who, coming as a waif and stray from Lincolnshire, could excite the admiration of such eminent people. Her friendship with Swift would suggest that she was no prude, and in his reports of their conversation – on one occasion dissecting the latest scandal and on another discussing the absence of virgins in London – she appears to have been spirited and lively enough to fascinate a man who was easily bored.

It seems that she was loved by more than Swift, however, and there are persistent reports that she became the mistress of Charles Montagu. Newton's erstwhile patron, who became Lord Halifax in 1700, was smitten by her – if the evidence of the surviving documents is anything to go by. When he drew up his will he bequeathed three thousand pounds and all of his jewellery to her as 'a small Token of the great Love and Affection I have long had for her', a bequest to which he added greatly in subsequent years when he bestowed upon her an annuity together with a great house and a manor.

There was of course much gossip about their relationship, all the more prurient because of the shadowy presence of the great Isaac Newton in the ménage. John Flamsteed, already smarting under Newton's treatment of his astronomical data, declared that Halifax had left her money and land 'for her *excellent conversation*', a satirical thrust which was not lost upon his contemporaries. Halifax's own biographer admitted the suspicions when he claimed that Halifax, on the death of his wife, had wished Catherine to become the 'Super-intendant of his domestick Affairs'; since she was 'young, beautiful, and gay' the censorious passed 'a Judgement upon her which she no Ways merited, since she was a Woman of strict Honour and Virtue'.

Voltaire, during his exile in England at a much later date, had heard a more scandalous version of events. 'I thought in my youth that Newton made his fortune by his merit', he wrote:

I supposed that the Court and city of London named him Master of the Mint by acclamation. No such thing. Isaac Newton had a very charming niece, Madame Conduitt, who made a conquest of the minister Halifax. Fluxions and gravitation would have been of no use without a pretty niece.

This is what might be called a Gallic interpretation of events. Newton had been appointed as Warden of the Mint before his niece had even arrived in London, and in the year he was appointed Master of that institution Halifax had fallen from influence. There was no way of preventing surmise, of course, and the fact that the story was still circulating at the time of Voltaire's visit in 1726 suggests how strongly embedded it had become. In a satirical tract of 1710, *Memoirs of Europe*, Halifax and Catherine appear under assumed names; the man has lavished goods on his mistress, 'besides getting her worthy ancient Parent a good Post for Connivance'. Halifax himself was not unaware of the irony of the situation. In a set of verses he wrote at the Kit-Cat Club in celebration of Catherine, he concluded with the couplet,

> Full fraught with beauty shall new flames impart,
> And mint her shining image on the heart.

The reference to 'mint' here was not accidental.

Newton must have been aware of the rumours, but there is no record of his reaction. It seems likely that in this matter he acknowledged the power of social mores, and made no objection to her liaison with the man who had materially advanced his prospects. Some commentators have cited his dissenting faith, and accused him of hypocrisy in sanctioning an immoral relationship. But he was in no respect a conventional man.

If she did indeed become the mistress of Halifax, there is no reason to believe that Newton disapproved of her conduct. She remained close to him for the rest of his life. In the one surviving letter to her, dated 5 August 1700, he signs himself 'Your very loving Unkle'. She had retreated to the country in order to recover from the smallpox and in his letter he hopes that the illness 'abates and that the remains of the smallpox are dropping off'. At the end of the letter he adds, 'Pray let me know by your next how your face [is] and if your fervour [fever] be going. Perhaps warm milk from the Cow may . . . abate it.' It is worth quoting this letter as one of the few examples of what might be called ordinary human feeling on Newton's part. On her part, in a letter that she sent to Newton on the death of Halifax, she signed herself 'your obedient niece and humble servant', which suggests a good measure of deference. After Halifax's death, she continued to be Newton's housekeeper.

On his arrival at Jermyn Street Newton had fitted up his new home in luxury, having a particular propensity for crimson furnishings. It is one of the strange aspects of his character that he was obsessed by the colour crimson. In the inventory of his possessions, drawn up after his death, there is reference to a 'crimson mohair bed' with 'case curtains of crimson', crimson drapes and crimson wall hangings, a crimson settee with crimson chairs and crimson cushions. There have been many explanations for this, including his study of optics, his preoccupation with alchemy, or his desire to assume a quasi-regal grandeur. But it may simply be a mark of his difference from the rest of the world, his uniqueness, a flash of his singular genius in an unexpected setting.

# Chapter Fifteen
# Leader of the Royal Society

He became part of the larger London world, having kissed hands with William III at the time of his appointment, but he did not partake of the usual London entertainments. He seems to have been averse to snuff and tobacco, the familiar recreations of the period, explaining that 'he would make no necessities to himself'. His food seems to have been plain, and his drink simple. He was not interested in the exhibitions of sculpture that were so fashionable in the period; he said of one connoisseur that he 'was a lover of stone Dolls', which evinces an almost biblical distaste. He rarely attended musical evenings, and described his only attendance at the opera as a mixed enjoyment – 'the first act, said he, I heard with pleasure, the second stretched my patience, at the third I ran away'. He seems not to have read any literature, and once described poetry as 'a kind of ingenious nonsense'.

Instead he seems to have spent his evenings in the city, when he was not entertaining his new set of London acquaintances, in his old habits of research and study. One contemporary reported that 'he was hardly ever alone without a pen in his hand & a book before him'. He was still deeply involved in his biblical and chronological studies, and never lost his interest in alchemical topics. There is a note in Newton's hand about a book entitled *Sanguis Naturae. Or, a Manifest Declaration of the Sanguine and Solar Congealed Liquor of Nature*. Since it was not published until after Newton's removal to London, it must be a late purchase. Newton wrote that this book and others could be procured 'at Sowles a Quaker Widdow in White Hart Court at the upper end of Lombard Street'.

He had not lost his touch for calculus, either. A mathematician from the University of Groningen, Johannes Bernoulli, set him a public challenge to resolve two very complicated calculations concerning the path of heavy bodies. Catherine Barton has told the story. 'When the problem in 1697 was sent by Bernoulli Sir I.N. was in the midst of the hurry of the great recoinage did not come home till four from the Tower very much tired, but did not sleep till he had solved it which was by 4 in the morning.' He achieved in twelve hours what most of his colleagues could not have achieved in twelve years. Bernoulli confessed himself humbled, and said that he had recognised Newton's solution 'as the lion is known from its claws'. This is an apt simile, suggesting the power and ferocity of Newton's mind.

His reputation was now known all over the European world. When the tsar of Russia, Peter I, arrived in London in 1698 one of his keenest expectations was of meeting Isaac Newton. He was inspecting the Mint as part of his visit, and Newton was informed that 'he likewise expects to see You here'. Newton duly obliged, and no doubt found the practical and ingenious tsar more knowledgeable about his theories than his own sovereign.

His position was in fact strengthened when, at the end of that year, he was appointed to be Master of the Mint. This was the position that he had always desired, and at last he had complete control of the institution to which he devoted much his working life. He was appointed master on 25 December, perhaps chosen for his birthday, and did not relinquish the position until his own death some twenty-seven years later. He at once cleansed the Augean stables of his predecessor's maladministration. Newton set up a proper system of accounting, and inaugurated a weekly meeting of the Mint board. He was also intent upon increasing the efficiency and reliability of the process of coining. On his own account he stated that 'I have brought the sizing of the gold & silver moneys to a much greater degree of exactness than ever was known before, & thereby saved some thousands of pounds to the government . . .'.

He was now an affluent man. His annual salary had risen to five hundred pounds but, more importantly, he received a profit on every pound weight of silver that was coined. His predecessor in the office had thereby acquired a fortune of £22,000 and there was no reason to believe that Newton would gain any less. That is perhaps why he resigned his academic posts two years after his appointment; the emolument was too small to be of any practical significance.

He had in any case other positions of responsibility. In 1701 he once again became a Member of Parliament for Cambridge University, and sat in the Commons for some eighteen months. He was not prominent in its deliberations, however, and his only known vote was in support of his erstwhile patron Halifax and other prominent Whigs. He was placed there as a party stalwart, and refused to stand openly in the election of 1702. 'Now I have served you in this Parliament,' he wrote to the vice-chancellor, 'other gentlemen may expect their turn in the next.' He also seems to have been constitutionally averse to opposition of any kind, and remarked that 'to solicit and miss for want of doing it sufficiently, would be a reflection upon me, and it's better to sit still'.

There were other avenues for his energy and ambition, however, and in the following year he was elected as President of the Royal Society. His attainment of that honour occurred during a year when the business of the Mint had slackened somewhat. There was no new coinage in the last eight months of 1703, and he needed a diversion. The timely death of his old enemy, Robert Hooke, Secretary of the Royal Society until his demise, smoothed Newton's path.

Newton took over the society at a lean time in its fortunes. Its original membership of two hundred had almost halved, it was facing financial insolvency and it was about to be evicted from its premises in Gresham College. Enter Isaac Newton. He retained the presidency until his death, and in the course of the next two decades he revolutionised the character and reputation of the Royal Society until it became the chief expression of scientific opinion in Europe. He was elected on the last day of November 1703, and when he attended his first meeting as president on 15 December it was clear that he was going to be a forceful and even wilful administrator.

Previous holders of the office, often aristocratic placemen like Halifax himself, rarely bothered to attend meetings; Newton changed all that. He was present at almost all of the meetings called during his lifetime, and presided over them in his characteristically attentive and authoritarian fashion. He often summarised the contents of the discussion, and then announced his own thoughts on the matter *ex cathedra* from the president's special chair at the head of the table. Only when he had settled himself in this particular chair was the official mace of the society placed on the table by a liveried servant. The mace was not employed unless and until Newton was present. The Society was in essence a kind of court with its own sovereign.

William Stukeley was a member of the society, and has left his own record of its proceedings under Newton's guidance. 'While he presided in the Royal Society,' Stukeley wrote, 'he executed that office with singular grace and dignity – conscious of what was due to so noble an Institution – what was expected from his character.' So the 'character' or public image of Newton had already become an emblem of his authority and his prestige. At these meetings 'there were no whispering, talking nor loud laughters. If discussion arose in any sort, he said, they tended to find out truth, but ought not to arise to any personality.' This was the new principle of objective and impersonal science that Newton, more than any other person, introduced into the world. It can be said that he created the role and the 'character' of the disciplined and dedicated scientist.

So Stukeley recalls that 'Every thing was transacted with great attention and solemnity and decency; nor were any papers which seemed to border on religion treated without proper respect'. Science itself was in the process of becoming a new form of religion, despite Newton's own deep piety, with its laws and principles treated as a new form of unassailable dogma. Again Newton himself was responsible for the sea-change in public attitudes. Stukeley goes on to note that 'his presence created a natural awe in the assembly; they appeared truly as a venerable *consessus Naturae Consiliariorum*, without any levity or indecorum'. If any member did introduce a note of 'levity' into these august proceedings, he was asked to leave the room.

Newton had also arrived with a 'scheme' in which he outlined the nature and purpose of the society's deliberations. He declared that natural philosophy 'consists in discovering the frame and operations of Nature, and reducing them, as far as may be, to general Rules or Laws – establishing these rules by observations and experiments, and thus deducing the causes and effects of things'. This became the working definition of the scientific method. He also laid down the principal areas of inquiry – arithmetic and mechanics, astronomy and optics, the 'philosophy relating to animals' with particular attention to anatomy, the 'philosophy relating to vegetables' which would now be called botany, and, finally, mineralogy and chemistry.

He may have wished to appoint demonstrators or curators in each of these disciplines, but the finances of the society did not encourage such expense. His first appointment, however, was a replacement for Robert Hooke as the secretary; his choice was Francis Hauksbee who, apparently at Newton's instigation, began a series of experiments on a new air-pump.

But the high scientific tone of the society's deliberations could not be continued indefinitely. It was still in part an association of amateurs because there was then no concept or definition of the professional 'scientist'. So there were disquisitions on the birth of a dog with no mouth, on the curative properties of 'cow's piss', and on the penis of a possum. When one foreign observer attended a meeting he was somewhat scornful of the proceedings as the work of 'apothecaries and the like'. Of Newton he remarked that he 'is an old man, and too much occupied as master of the mint, with his own affairs, to trouble himself much about the society'.

This was not strictly true, and in the months following his appointment he presented his long-awaited study of optics to the Royal Society. Now that his old rival, Hooke, was safely in his grave he felt able to proceed. To the presentation he added an 'Advertisement' in which he explained that he had suppressed his thesis, from its inception in 1675, 'to avoid being engaged in Disputes'. Edmond Halley was chosen to read it for the society, and to provide them with a summary of its contents and conclusions.

Newton began with a declaration of his intent. 'My Design in this Book', he wrote, 'is not to explain the Properties of Light by Hypothesis, but to propose and prove them by Reason and Experiment.' He was setting out his vision of experimental science, therefore, as an alternative to the hypotheses and theoretical principles of the natural philosophers who had considered the phenomenon of light in the past. He had also written the book in English, unlike the Latin of *Principia Mathematica*, in order that he might reach a wider audience of his countrymen.

The text was indeed filled with experiments and observations, divided into definitions and 'quaeries', axioms and propositions. In his first experiment Newton describes how he took 'a black oblong stiff Paper' and directed light upon it from a prism in a darkened room. In a series of meticulous and detailed experiments he then demonstrated that 'Colours are not *Qualifications of Light* . . . but

*Original* and *connate properties*' and that 'Whiteness' is not pure light but heterogeneous and 'ever compounded'. These were conclusions which he had reached many years before, but which were now placed before an amazed audience of readers. Colours were not secondary aspects or properties of light; they were an integral part of divine creation and *were* light. He listed the seven colours of the spectrum – red, orange, yellow, green, blue, indigo and violet – on the analogy with the seven notes of the musical scale. There were originally thought to have been only three colours of the spectrum, but Newton insisted that there must be seven in accordance with Pythagorean principles of harmony. This may be considered a mathematical, or a mystical, vision of the universe.

He speculated on the corpuscular nature of light, also, but for him it was an hypothesis that could not be proved. He relegated it to a query in a revised Latin version of the *Opticks*, published in 1706, in which the question is posed: 'Are not the rays of light very small particles emitted from shining substances?' Some of his disciples were less restrained, however, and a decade later one of them, George Cheyne, stated explicitly that 'Light is a Body, or material substance'.

Newton also implicitly placed his observations within the framework of his theories of gravitation, when he suggested that 'the parts of Bodies do act upon Light at a distance'. This notion of action 'at a distance', itself still a mysterious and unusual concept to his contemporaries, was emphasised in the first 'query' that he added to the first edition of *Opticks*. 'Do not Bodies act upon Light at a distance; and by their action bend its Rays; and is not this action (*ceteris paribus*) strongest at the least distance.' There are suggestions here of a unified theory that he was never finally able to formulate.

Newton stated that he had written out his experiments in English so that 'the novice [might] the more easily try them', and indeed the practical scope of Newton's book appealed to the empirical temper of the English. The *Opticks* had in fact a greater short-term impact than the *Principia*. His method of deriving laws and principles from multifarious experimentation became the model of scientific inquiry itself. In one of his epistles from England Voltaire praised Newton as one who had been able to 'anatomize a single Ray of Light with more Dexterity than the ablest Artist dissects a human body'. The famous statue of Newton, fashioned by Roubiliac, shows him with a prism. The Romantic poets of England particularly attacked his analysis of light as a token of the moral and aesthetic feebleness of scientific inquiry; in *Lamia* Keats mourns the fact that

> Philosophy will clip an Angel's wings,
> Conquer all mysteries by rule and line . . .
> Unweave a rainbow.

It is clear that the object of his scorn here is Isaac Newton.

## Chapter Sixteen
# A battle of wills

In the early spring of 1704, a few months after he had been elected as President of the Royal Society, Newton once more paid a call upon John Flamsteed, Astronomer Royal, at the Greenwich Observatory. He was in fact the official astronomer of a new sovereign, Queen Anne having assumed the throne two years before. Newton visited him ostensibly to check progress upon Flamsteed's 'star catalogue' and to recommend its patronage to Prince George, the husband of the new queen. But essentially he wanted to use the astronomer's observations in order to make progress with his own lunar theory. The meeting was not wholly successful. When Flamsteed questioned a supposed fault in the *Principia* Newton asked: 'Why I did not hold my tongue?' The flash of anger is indicative of Newton, defensive and querulous when he believed himself to be criticised.

Flamsteed proceeded to thank Newton for sending him a copy of *Opticks*. 'He said then he hoped I approved it', Flamsteed wrote later. 'I told him loudly "no", for it gave all the fixed stars bodies of 5 or 6 seconds diameter, whereas four parts in five of them were not 1 second broad. This point would not bear discussion; he dropped it, and told me he came now to see what forwardness I was in . . .' But Flamsteed would not drop his criticisms of Newton. 'I

showed him also my new lunar numbers fitted to his corrections, and how much they erred, at which he seemed
surprised, and said "it could not be".' As he left the
observatory he told Flamsteed to 'do all the good in your
power' – to get on with his catalogue, in other words.
Flamsteed noted later that doing good was the principle of
his life, 'though I do not know that it has ever been of his'.
This note, of recrimination and suspicion, was one that
endured during their encounters. In a later memorandum
Flamsteed noted of Newton that 'I know his temper; that he
would be my friend no further than to serve his own ends'.
He also concluded that he was 'spiteful, and swayed by those
who were worse than himself'. These remarks were written
in 1717 in Flamsteed's preface to the eventual publication of
his work. But the preface was suppressed. It was not
considered politic, or respectable, to criticise the grand old
man of English science.

In turn Newton was not fond of Flamsteed; he was not
fond of anyone who questioned his judgements. But he
needed the star catalogue to continue his own calculations.
So he proceeded to circumvent the astronomer in order to
acquire his observations. Flamsteed was not invited to join
the delegation that solicited the patronage of Prince
George. Flamsteed was not one of those chosen to supervise
the publication of the catalogue, although he had drawn up
his own plans for just such publication. When the
astronomer wrote requesting a new translation of Ptolemy's
star catalogue, Newton did not reply to him. Instead
Flamsteed was to be granted £180 for assistance to help him
calculate 'the places of the moon and planets & comets'.
This was the only information that Newton required.

Flamsteed realised soon enough that he was being excluded from all deliberations concerning his own star catalogue. 'I do not remember that I was present at any more of their meetings but one,' he wrote, 'where nothing material was determined whilst I was present; though I considered that [I was] the person chiefly concerned . . .' He was incensed, too, that the putative publisher was to be paid thirty-four shillings per sheet of the catalogue whereas Flamsteed himself was not to be reimbursed at all. He had devoted thirty years of his life to his observations, and was expected to be content with his annual salary of £100 as Astronomer Royal. Out of that salary, as he never tired of pointing out, he had to purchase his own equipment and hire his own assistants. He had asked the committee, established by Newton, to procure 'an honorable recompense for my paines, and *2000lib.* [pounds] in expense'. The committee refused to consider it.

It is clear enough that Flamsteed was very badly treated by Newton, who seems to have considered him as at best a nuisance and at worst a necessary evil. Flamsteed knew as much himself, and came to realise that Newton wanted the star catalogue for his own purposes. He noted in a memorandum that 'Sr I N having been furnished from My Stores would have me thrash it all out my selfe & charitably bestow it on the publick that he may have the prayse of haveing procured it'. He wished to be recompensed for the sake of his reputation as much as his pocket, since 'my Countries Ingratitude would be attributed, by Sir I N himselfe, to my *Stupidity*'.

Negotiations over the publication of the star catalogue continued throughout 1704 and 1705, much to Newton's annoyance. There were disputes over errors of observation, over details of publication, and over problems of manuscript delivery. Flamsteed noted of Newton that 'I doe not court him, & his temper wants to be cried up & flattered'. Eventually, in March 1706, Flamsteed came to Newton's house where all parties reached agreement. 'Sir Issack askt me if things went not now to my Content I returnd that it was strang that I should be so little taken notice of who was the person mainly concernd at which he seemed chagrin.'

Flamsteed delayed transmitting to the publisher the catalogue of the fixed stars, for the principal reason that it was not yet ready for the press, and this instigated further confrontation with Newton himself. There seems to have been some explosion of rage, since Flamsteed noted Newton's 'proud and insolent temper'. He went on to report that 'he has been told calmly of his faults, and could not contain himself when he heard of them'. This sounds very like Newton in a fit of rage against the man who dared to question and challenge him. As Flamsteed added, 'I always found him insidious, ambitious and excessively covetous of praise, and impatient of contradiction'.

Yet there were still delays, disputes and misunderstandings so that at the time of Prince George's death, in 1708, the project was put in abeyance for want of a suitable patron. Newton was furious at being denied the opportunity of studying Flamsteed's observations, and in the following year he had the Astronomer Royal removed from the membership of the Royal Society for non-payment of his subscription. Flamsteed wrote to a colleague that the society was being 'ruined' by Newton's 'close, politic and cunning' behaviour. There would be further eruptions.

Flamsteed did note a meeting in 1705, however, when Newton seemed 'more than usually gay and cheerful'. This was in the period when he had been awarded a knighthood by Queen Anne. He had not been honoured for his services to science, or to the Mint, but for his political services to the nation. Nevertheless he was the first natural philosopher and mathematician to be so honoured. The queen had travelled to Cambridge for the ceremony, and Stukeley recalled that 'her majesty dined at Trinity college, where she knighted Sir Isaac, and afterward, went to Evening Service at King's college chapel'. After the ceremony the students stood close by 'as he sat with heads of the colleges; we gaz'd on him, never enough satisfy'd, as on someone divine'. His new honour did not assist his political career, however. When he stood in this year as Member of Parliament for the university, he came last in the ballot. His parliamentary life was over.

Yet he maintained his hold upon the Mint, in a period when recoinage in Scotland and fresh outbreak of war on continental Europe placed much strain upon the orderly workings of the presses. He could not entirely break with his learned past, however, and continued with his researches into alchemy and theology; these were the two branches of learning he had not fully conquered. On the backs of records from the Mint, Newton wrote down his notes on ancient religions and biblical revelation.

His own beliefs were confronted when, in 1707, his friend Fatio de Duillier enrolled himself with the fanatical French sect of the Camisards. They had arrived in London, and were preaching the fulfilment of Revelation and the destruction of the Anti-Christ of Rome. Newton had an instinctive sympathy with their millennial ultra-Protestantism, and one contemporary recalled that 'Sir Isaac himself had a strong inclination to go and hear these prophets, and was restrained from it, with difficulty, by some of his friends, who feared he might be infected by them as Fatio had been'. But their prophecies of a second Great Fire of London, and of the combustion of the Lord Chief Justice, were not to the taste of the authorities and the French prophets were placed in the stocks. Newton was too aware of his position in public society to allow himself to become associated with them. Fatio himself was placed in the pillory, and there is no record of any further communication with Newton.

There were some aspects of his learning that were less controversial. He arranged a Latin version of *Opticks*, and had asked a young disciple, Abraham de Moivre, to superintend its publication. De Moivre later recalled how Newton would wait for him at Slaughters Coffee-House, on St Martin's Lane, and how they would then adjourn to Newton's house close by to discuss philosophical matters. Newton added certain further 'queries' to the Latin edition in which he speculated more openly about the constituents of the universe. In 'query 31', for example, he hinted at a 'unified' theory in which 'the Attractions of Gravity, Magnetism and Electricity' are all related.

In 1706 his lectures, as Lucasian professor, were published under the title of *Arithmetica Universalis*; they were published anonymously but there were so many mistakes within the text that Newton felt obliged to prepare another edition some sixteen years later. He was always revising, and extending, his legacy. He began work, for example, on a second edition of *Principia Mathematica*. It was eventually published in 1713, and proved to be in great demand. The *Principia* has never since gone out of print.

A young mathematician, Roger Cotes, was chosen to supervise the new edition; he had the courage, or temerity, to point out many mistakes in Newton's calculations. Newton grudgingly assented to the alterations, and in fact entered a correspondence with Cotes that was one of the most fruitful of his life. Yet in the next edition Newton chose not to acknowledge Cotes's contribution. It is another example of his endless demand for power and control.

He also tightened his grip upon the Royal Society. He was instrumental in the expulsion of one member, John Woodward, from the council on the grounds of unseemly conduct. He succeeded in his aim, and then put the mark of an 'X' beside the names of all those who had supported Woodward. He wanted them to be defeated at the next council election.

He was also immensely high-handed in his choice of new premises for the society. Woodward had granted the society permission to use his apartments in Gresham College for their meetings but, on his removal from the council, it was decided by Newton that it would be impolitic to rely on Woodward's generosity. Newton therefore set about finding new quarters. By the early autumn of 1710 he revealed at a special meeting that a house in Crane Court, off Fleet Street, was up for sale. A committee was set up to examine the matter, and Sir Christopher Wren was sent to survey the building.

At one meeting of the council some members complained that they had been given no notice of the removal from

Gresham College, and indeed doubted that it was necessary. According to one report Newton then remarked that he was not 'prepar'd . . . to enter that Debate; But freely (tho' methinks not very civilly) reply'd, That he had good reasons for their Removing, which he did not think proper to be given there'. He was asked why he had even summoned them. At which point he adjourned the meeting. These were the actions of the autocrat. The house in Crane Court was of course purchased.

He was involved in transferring himself during the same period. In 1710 he moved to a house in St Martin's Street, a few yards south of Leicester Fields (the site of the house is now sandwiched between Leicester Square and the back of the National Gallery). He had originally moved from Jermyn Street to Chelsea in 1709, but the air or neighbourhood of that riparian borough did not appeal to him. Nine months later he moved on to 35 St Martin's Street, where he remained for the next fifteen years. It was of three storeys, with a basement, and the *Survey of London* reveals that Newton built an observatory upon the top floor.

In new 'Orders of the Council', promulgated after the removal of the Royal Society to Crane Court, it was duly decreed that only Newton might sit at the head of the table and that members were only allowed to speak on being addressed by Newton himself. Since he was obliged to attend Mint business on Wednesday afternoons, it was agreed that meetings of the Royal Society should take place each week on the day following. Newton's authority was complete, and he was celebrated by one member as the 'perpetual dictator' of the society.

The portraits of him in this year, at the age of sixty-seven, show him without a wig; his expression is still one of alertness and eagerness, with the slightest trace of self-satisfaction. He had become the grand autocrat of science. He was asked to represent scientific opinion on various administrative committees. He sat on the parliamentary committee, for example, chosen to determine the means of measuring longitude at sea. He institutionalised science, with himself as its titular head. He came to represent the English genius, marked by utility and practicality, and could even be described as the most potent symbol of Western science itself. He was sometimes called 'the divine Newton' in honour of his status.

# Chapter Seventeen
# Duel of wits

Even though his authority was now complete, it was still open to challenge from contemporaries both at home and abroad. One of the fiercest of his antagonists was still John Flamsteed. The plans for publishing his star catalogue had been put in abeyance after the death of Prince George in 1708, much to the frustration and annoyance of Newton who needed the astronomer's calculations. In the two succeeding years Flamsteed had finally completed his catalogue. What could be more coincidental, therefore, than at the end of 1710 Queen Anne issued a warrant naming Isaac Newton and other members of the Royal Society as 'perpetual Visitors' at the Greenwich Observatory, permitting them to investigate all aspects of the Astronomer Royal's work and to scrutinise all of his observations? Flamsteed always believed that Isaac Newton was instrumental in setting up this warrant and others like it, and there is little reason to disagree with him.

John Arbuthnot, a member of the Royal Society and physician to the queen, then wrote to Flamsteed commanding him to deliver the star catalogue. Flamsteed replied that he needed assistance to complete that work. The next letter came from Newton himself. 'I understand', he wrote in terms of repressed fury, 'that . . . you have given an indirect & dilatory answer.'

He reminded Flamsteed that 'the observatory was founded to the intent that a complete catalogue of the fixt stars should be composed by observations to be made at Greenwich & the duty of your place is to furnish such observations'. Flamsteed was thereupon 'desired' to send the catalogue at once so that work on its publication could resume, and 'if instead thereof you propose any thing else or make any excuses or unnecessary delays it will be taken as an indirect refusal to comply with Her Majesties order. Your speedy & direct answer & compliance is expected.' It might seem that Flamsteed could expect no less than beheading at the Tower if he refused. He would certainly have feared for his employment.

Flamsteed met Arbuthnot at a coffee-house, and agreed to deliver the rest of the catalogue. But then he learnt that corrections and alterations were being made in his text without his approval. He wrote a long and pained letter to Arbuthnot in which he asked: 'would you like to have all your Labours surreptitiously forced out of your hands, convey'd into the hands of your declared profligate Enemys, printed without your consent . . .?' This of course meant nothing to Newton, who went ahead with the publication of those parts of the catalogue that were of most use to him.

He took a more exacting vengeance. He ordered Flamsteed to appear before him at Crane Court, in order to report on the state of his astronomical instruments. This was the confrontation that Flamsteed required. He arrived at the meeting, and promptly told Newton that the equipment was of his own purchase and therefore beyond Newton's control. At this point, according to Flamsteed's report to a friend, Newton 'ran himself into a great heat & very indecent passion'. 'As good have no Observatory as no Instruments', Newton is supposed to have replied. That may have been considered a veiled threat.

Then Flamsteed went back to the burning subject. He complained vehemently about the publication of his star catalogue. 'At this he [Newton] fired & cald me all the ill names Puppy etc that he could think of.' It seems clear that Newton had a short temper. Flamsteed 'put him in mind of his passion desired him to govern it & keep his temper. This made him rage worse . . .' Newton then reminded Flamsteed of his annual salary from the government, to which the astronomer replied in a scathing fashion. 'I asked what he had done for the 500 lb per Annum that he had received ever since he setled in London . . .' It was not a happy encounter; Flamsteed eventually took his leave and departed. The astronomer did eventually publish his own version of the star catalogue, having burnt all the earlier copies he could find, so that he might be said to have had the last word.

The imbroglio does not place Newton himself in the most favourable light, however. He showed himself to be ruthless and unforgiving, fierce and aggressive when crossed, prone to anger and intolerance. There is a note of cruelty in his temperament, too, associated both with his competitiveness and with his need to dominate and to control. These unhappy characteristics cannot be altogether separated from his desire to impose order upon the forces of the universe. The visionary was also the autocrat.

Soon enough he found another antagonist further afield, when he became entangled with the German mathematician and philosopher, Gottfried Wilhelm Leibniz. Both of them had claimed priority in the invention of calculus, or what was known as 'the arithmetic of fluxions', and each accused the other of plagiarism. In the spring of 1711 Leibniz had sent a letter to the secretary of the Royal Society, Hans Sloane, in which he attacked various parties 'for having attributed to myself another's discovery'. He was complaining, in other words, that Newtonian disciples had publicly hailed Newton as the 'first inventor' of the 'now highly celebrated arithmetic of fluxions' – one of them adding that 'the same arithmetic . . . was afterwards published by Mr Liebnitz in the Acta Eruditorum'. The charge against Leibniz was that of intellectual theft.

In turn, in a series of anonymous papers, Leibniz had himself accused Newton of plagiarism. As Newton put it, 'these papers everywhere insinuate . . . that the method of fluxions is the differential method of Mr Leibnitz & do it in such a manner as if he was the true author & I had taken it from him'. The objective truth of the matter seems to be that both men had formulated approximately the same method independently, in one of those acts of simultaneous genius that are relatively common in the history of science. What is indubitable is that Leibniz published his formulation first. But in the dispute between Newton and Leibniz the objective truth was not an issue. It became a cat-fight, both men snarling and hissing and clawing in the struggle for pre-eminence. They were both men of genius – Leibniz has been described as 'one of the greatest polymaths in history' – but in this matter they behaved like children.

After some acrimonious correspondence Leibniz asked the Royal Society, at the beginning of 1712, to adjudicate on the matter. He could not have made a bigger mistake. Newton declared that the members of the committee, set up to investigate the question of priority, were 'numerous and skilful and composed of Gentlemen of several nations, and the Society are satisfied in their Fidelity'. In reality they were all chosen by Newton himself. Newton collated all the documents, marshalled all the evidence and even wrote the final report in his own handwriting. Never had his power been wielded so shamelessly. So the committee reported, without any dissenting voices, that 'we reckon Mr Newton the first Inventor'. Newton later had the gall to reflect, in connection with Leibniz, that 'no Man is a witness to his own Cause'. Newton had not only been witness; he had also been judge and jury of his own case.

A version of the judgement was printed and distributed to the universities and other centres of learning. Newton's victory was complete. In a draft preface to the published judgement, entitled *Commercium Epistolicum*, he wrote that 'The first inventor is the inventor & whether the second inventor found it by himself or not is a question of no moment'. Or, as he put it more plainly in a note, 'Second inventors count for nothing!' He removed the name of Leibniz from the third edition of the *Principia*, a symbolic act of erasure that had already been meted out to Hooke and Flamsteed.

He told a friend, after Leibniz's death, that he 'had broke Liebnitz's Heart with his Reply to him'. The pursuit of science had no moral, or ethical, dimension. Newton never once questioned his own rightness or doubted his own motives. It really did seem that he considered himself to be God's appointed, beyond reproof or reproach. He was benign as long as he was worshipped, but the unbelievers were cast into outer darkness.

It seems that the heart of Leibniz did not break, however, under the strain of Newton's contempt. In 1713, under the cloak of anonymity, he wrote a treatise in reply to *Commercium Epistolicum*, entitled *Charta Volans* or 'Flying Sheet', in which he once more accused Newton of blatant theft. He said that Newton 'took to himself the honour due to another of the analytical discovery or differential calculus first discovered by Leibniz . . . he was too much influenced by flatterers ignorant of the earlier course of events and by a desire for renown'. This was wounding enough but then Leibniz added that 'of this Hooke too has complained, in relation to the hypothesis of the planets, and Flamsteed because of the use of his observations'. Leibniz was very well informed of the internecine warfare between the English natural philosophers. Newton annotated the treatise furiously.

The dispute spread to the pages of learned European periodicals and, in the manner of such controversies, enmired many other people. The news of the fight circulated through the court; the new king of England, George I, had been an employer and patron of Leibniz while Elector of Hanover. In one letter Newton remarked that 'I was pressed for an answer to be also shewed to His Majesty, & the same afterwards sent to Leibnitz'. It is rare that abstruse matters of mathematical calculation become the object of royal attention. It became almost a duel between two rival nationalities, with the 'English' calculus of Newton being set against Leibniz's more accessible and comprehensible version. As it turned out, the mathematicians of Europe did indeed favour Leibniz.

# Chapter Eighteen
# In decline

In the summer of 1717 Newton's niece and house-keeper, Catherine Barton, became engaged to John Conduitt, some nine years her junior; Conduitt was handsome, clever and rich which, as the readers of Jane Austen know, is a heady combination. He had acted in various capacities for the British army in Portugal and Gibraltar, but he was also an amateur archaeologist. He had located and identified the ancient Roman city of Carteia, in Gibraltar, and had given a paper to the Royal Society on that topic. It may have been in this capacity that he and Newton first met. The uncle led to the niece. In the manner of business of those days he took up a position at the Mint, and eventually succeeded Newton as its Master.

Conduitt remained in awe of Newton for the rest of his life. He kept a memoir of him and, after his death, went to great lengths to preserve his legacy. His anecdotes of Newton represent some of the few first-hand reports of Newton in his last years, even if they seem somewhat softened by hero-worship. He described his existence as 'one continued series of labour, patience, humility, temperance, meekness, humanity, beneficence & piety'. This is not a description that would readily have occurred to Leibniz or Flamsteed. Conduitt also described Newton as possessing a full head of white hair, and the complexion of a much younger man with 'a very lively piercing eye'.

There were other accounts of Newton in the last years of his life. Bishop Atterbury, a contemporary acquaintance, remarked that in his features 'there was nothing of that penetrating sagacity' to be found in his compositions; indeed Newton 'had something rather languid in his look and manner'. Another contemporary, Thomas Hearne, agreed that he was 'of no promising Aspect'. He was short, or what others called 'of middling height', and 'he was full of thought, and spoke very little in company, so that his conversation was not agreeable'. This was a significant defect, in an age when the art of conversation was considered indispensable in the business of life. Hearne also added the odd detail that 'when he rode in his Coach, one arm would be out of the Coach on one side, and the other on the other'. It is not clear what this signifies except, perhaps, a certain lordliness in his demeanour.

His habits were temperate in old age, with a breakfast of bread and butter with orange tea. He drank water principally, and had some of the instincts of a vegetarian. He refused to eat black pudding because it was made with blood, and would not eat rabbits because they had been strangled. It is also worth mentioning here that he was well known for his charity; he helped various members of his own family, but he also distributed money to strangers who wrote him 'begging letters'. This charitableness is of course consonant with his Christian beliefs, and he was now a rich man, but it helps to modify the impression of a distant and authoritarian figure.

Newton's tenant at Woolsthorpe, Mr Percival, confirms Hearne's impression of his reticence. He recalled that

Newton 'would sometimes be silent and thoughtful for above a quarter of an hour together, and look all the while almost as if he was saying his prayers'. But, when he did speak, it 'was always very much to the purpose'. This seems authentic – a taciturn and reflective man who spoke to the point.

William Stukeley, an admirer, later wrote a biographical memoir of Newton, in which he stated that he 'was of a very serious and compos'd frame of mind, yet I have often seen him laugh . . . He had a good many sayings, bordering on joke and wit. In company he behavd very agreeably; courteous, affable, he was easily made to smile, if not to laugh.' This in itself sounds somewhat forbidding, and there is no doubt that Newton remained a formidable and on occasions somewhat frightening figure.

He continued his business at the Mint with his old vigour and in the autumn of 1718, in his seventy-ninth year, wrote a set of 'Observations upon the State of the Coins of Gold and Silver'. It was he who was instrumental in fixing the value of the guinea at twenty-one shillings, a measure that remained in force for almost three hundred years. In the year before he had assumed responsibility for the coinage of copper halfpennies and farthings, and had created the means of testing the purity of the metal. His early years in the alchemical laboratory were still of service to him. He was able to assay, and reject, the supplies of copper from several different bidders. He had not lost his acerbity in matters of business. When one speculator announced that he had a method of coining which would preclude counterfeiting, Newton reported that 'I take him to be a trifler, more fit to emboyle the coinage than to mend it'. He also advised against clemency for a counterfeiter convicted to hang; it was, he wrote, 'better to let him suffer' than to permit him to teach others the tricks of his trade.

He also remained firmly in control of the Royal Society, presiding over its meetings with his usual vigour and regularity. He helped to subsidise its activities, and acted as a patron for the experiments of the young natural philosophers, for whom he was now the object of veneration. But, more significantly, he was engaged in preserving and extending his legacy. He published a second Latin edition, and a third English edition, of his *Opticks*. In 1722 he chose a young mathematician, Henry Pemberton, to superintend the third revision of *Principia Mathematica*. Pemberton later recalled that 'though his memory was much decayed, I found he perfectly understood his own writings, contrary to what I had frequently heard in discourse from many persons'. So Newton was rumoured to be of failing mind. Pemberton also noted that 'neither his extreme great age, nor his universal reputation had rendered him stiff in opinion, or in any degree elated'. He remained calm, and clear, until the end.

Despite his clarity of mind, however, he seemed no longer to have taken any delight in mathematics. He told a fellow mathematician that 'I never applied my self much to practical mathematicks, and they are now much out of my mind'. He informed another correspondent that 'its now above eighteen years since I left off the study of Mathematicks & the disuse of thinking upon these things makes it difficult for me to take them into consideration'.

But he had never stopped thinking upon matters of theology and biblical chronology. He was engaged in the study of biblical prophecy, also, with particular attention to the Book of Daniel. He seems to have come close to the opinion that the universe itself was the 'spiritual body' of Jesus Christ, subservient to the Father but instrumental in the making of the cosmos. This was aligned to his supposition that there was some universal 'ether' or spirit that created such forces as gravity. The 'spiritual body' of Christ was in some sense the living cosmos. He was getting closer and closer to the heart of the mystery, but he must have realised in the end that it was beyond the power of human intelligence to venture much further. He remarked upon his wish to 'have another touch at metals' and 'another shake at the moon', but he knew well enough that he was no longer capable of his earlier feats of calculation and creation.

At the close of his life he relayed a judgement of his own career. 'I don't know what I may seem to the world,' he said, 'but, as to myself, I seem to have been only like a boy playing on the sea shore, and diverting myself in now and then finding a smoother pebble or a prettier shell than ordinary, whilst the great ocean of truth lay all undiscovered before me.' Those words have been endlessly quoted, in large part because they seem to represent the limits of human wisdom and human enterprise.

This was not of course how his contemporaries, or indeed later generations, would have described his achievement. John Aubrey believed that the *Principia* represented 'the greatest highth of Knowledge that humane nature has yet arrived to', rivalling the claims of Plato and Aristotle and Galileo. Newton was considered to be a genius, a prophet, a visionary; indeed he has some claim to all of these epithets. He had formulated the proper nature of the scientific enterprise, by aligning inductive reasoning with mathematics and rigorous experiment. He had revolutionised the study of optics, and established the principles of celestial mechanics. With his discovery of the universal laws of gravity, he had rendered the invisible visible. He explained for the first time the nature of tides. He discovered the infinitesimal calculus. He invented a new system of chronology that, according to Gibbon, 'would alone be sufficient to assure him immortality'.

It is sometimes said that he was the one man who led the way towards the Industrial Revolution and the current ventures of space exploration. Newtonianism has not essentially been superseded by relativity or by quantum theory. Newton created a system of the universe – of force and inertia and mass, of action and reaction – that remains unsurpassed in its reliability and efficiency. Two contemporary mathematicians, Stephen Hawking and Werner Israel, have stated that 'Newton's theories will never be outmoded'. That is why Keynes described him as 'the last of the magicians'. Albert Einstein has celebrated him as 'the experimenter, the theorist, the mechanic and, not least, the artist in exposition'. Yet the paradox remains that this obsessive and occluded man, this alchemist and devotee of an heretical faith, has become an icon of rational science and of reason itself. As Pope put it:

Nature and Nature's laws lay hid in night;
God said, *Let Newton be!* And All was *Light*.

## Chapter Nineteen
# The last days

In the spring of 1722 he became ill, from the effect of kidney stones, presaging a decline of his health over the remaining five years of his life. It was after this onset of illness that he chose one of his protégés, Henry Pemberton, to be his editor. In the following year he became more seriously ill, and placed himself under the care of two doctors. He suffered from what Conduitt described as 'relaxation of the sphincter of the bladder' which caused him to urinate frequently and excessively. He was placed on a diet of fruit and vegetables, 'which he ate very heartily'. It was recommended that he use a sedan-chair for travel, but he insisted upon walking wherever possible. He had a phrase, 'use legs and have legs', that sounds like a saying from childhood. He was now eighty-one years old. Conduitt reports that in 1724 he 'voided, without any pain, a stone about the bigness of a pea, which came away in two pieces; one at some distance from another'. Even the most gruesome detail was considered worthy of record.

He was perturbed in another sense when in the same year a version of his biblical chronologies was published in Paris in unofficial form. He had prepared an 'abstract' or 'short chronology' for the princess of Wales, and it was this summary that arrived by indirect means into the hands of a French bookseller. To compound the offence this abstract –

or *Abrégé de la chronologie*, as it was entitled – was published
with a series of 'Observations' that listed Newton's apparent
errors. He was outraged and, in his mid-eighties, he wrote
seven drafts of a long essay refuting the criticisms. He had
lost none of his acerbity. The author of the 'Observations'
'hath undertaken to translate and to confute a Paper which
he did not understand, and been zealous to print it without
my Consent'.

He was so roused by the incident that he began to prepare
a complete edition of his chronologies, but he was still
revising it at the time of his death. The rector of St Martin-
in-the-Fields, Zachery Pearce, visited Newton a few days
before that event. He recalled that 'I found him writing over
his Chronology of Ancient Kingdoms, without the help of
spectacles, at the greatest distance of the room from the
windows, and with a parcel of books on the table, casting a
shade upon the paper'. 'Sir,' Pearce said, 'you seem to be
writing in a place where you cannot so well see.' Newton
replied that 'A little light serves me.' The writer of *Opticks*,
who had always been fascinated by the workings of the eye,
had retained his sight in extreme old age. Pearce went on to
record that 'he read me two or three sheets of what he had
written . . . I believe that he continued reading to me, and
talking about what he had read, for near an hour, before the
dinner was brought up'.

On his eighty-third birthday, Christmas Day 1725, he
showed William Stukeley his drawing of Solomon's Temple;
he believed that '. . . the *Divine* lays his mysterious plan of
future things in the scenes of the Jewish temple and service'.
From Newton's published and unpublished writings it
seems clear that he possessed some vision of a primitive

Church established by the sons of Noah; its principles were, simply, to love God and to love your neighbour. This first and best religion had been passed by the Hebrew patriarchs to the Israelites, and by Pythagoras to the Greeks and Egyptians. It was the foundation faith of which true Christianity, or Arianism, was the survivor.

Biblical prophecy and chronology were thus the abiding preoccupations of his old age. He was seeking to resolve the mysteries of the ancient, as well as the modern, world. Yet he never completed the task. His own chronology had intervened.

At the beginning of 1725, after suffering from a violent cough and inflammation of the lungs, Newton had been persuaded to move from St Martin's Street to the more healthy air of Kensington. He suffered here from an attack of gout but at his new address, between Church Street and Kensington High Street, he seemed to be improving in health. Conduitt recalls, however, that 'though he found the greatest benefit from rest, and the air at Kensington, and was always the worst for leaving it, no methods that were used could keep him from coming sometimes to town'. The final phrase, 'without any real call', was subsequently deleted. He could not keep himself from activity and business in the world. He would arrive at the Mint or at the Royal Society unannounced, like some monarch secretly surveying his realm. Yet when he took a distinguished French visitor, the Abbé Alari, to a meeting of the society he began to doze during disquisitions on French bottles and Swiss weather. In the last portraits to which he sat, in 1725 and 1726, there seems to be a change. They were both executed by John Vanderbank but in the first of them, he looks remarkably sturdy and resolute; in the second the artist provides small intimations of age and debility. The gaze is more wayward, the visage less well defined.

In the last full year of his life, under the constant ministrations of his doctors, he maintained the semblance of activity. There was some correspondence concerning the third edition of *Principia Mathematica*, which was published in this year, and in the autumn he travelled to the Royal Mint in order to examine a new supply of gold ingots. He still attended the meetings of the Royal Society, but not as regularly as before. He also began to apportion his considerable estate.

But he was now concerned with the evidence of the past. It was reported by John Conduitt that he burned a number of papers in anticipation of his demise. Conduitt described the material as 'boxfuls of informations', and another witness called them 'manuscripts', but the content of these papers is unknown. Conjecture varies from the destruction of family letters to the burning of his heretical speculations. There has even been a suggestion that he wished to erase any evidence of his involvement in the black arts.

On 28 February 1727, he travelled east to London from Kensington; then, on 2 March, he presided over his last meeting of the Royal Society. On the following day Conduitt told him that he had not seen him so well for many years. Newton smiled and told his nephew that 'he had slept the Sunday before from eleven at night till eight in the morning without waking'. Yet appearances were deceptive. On Friday 3 March, he became violently ill with his old complaint of the sphincter; he returned to Kensington, but there was no improvement. He remained in pain for a week and, when the doctors examined him on Saturday 11 March, they surmised that the stone had 'probably been moved from the place where it lay quiet by the great motion and fatigue of his last journey to London'. It was now lodged in his bladder, and they held out no hope for his recovery.

He was enduring spasms of great pain, and Conduitt noted that 'though the drops of sweat ran down his face, he never complained, or cried out, or showed the least signs of peevishness or impatience'. In fact 'during the short intervals from that violent torture, would smile, and talk with his usual cheefulness'. William Stukeley reported that the pain 'rose to such a height that the bed under him, and the very room, shook with his agonys'.

Even in this extremity he refused the sacrament of the established Church. He would not subscribe to the heresy of trinitarianism a few hours before meeting the creator of the universe. He seemed to be gaining strength and, on the following Saturday, he read the newspapers. But that evening he slipped into a coma, from which he never awoke. He died in the early hours of the morning of Monday, 20 March 1727. The minute book of the Royal Society recorded his passing in cryptic terms. 'The Chair being Vacant by the Death of Sir Isaac Newton there was no Meeting this Day.' His body lay in state in the Jerusalem Chamber beside Westminster Abbey, before being carried to its burial place in the nave.

Sir Isaac Newton left an estate of £31,821, a grand sum for the son of humble yeomen, but he left no will. Yet he did not need to do so. The world had already been given its inheritance.

# Select Bibliography

*Primary Works*

Isaac Newton, *The Principia, Mathematical Principles of Natural Philosophy*, newly translated by I. Bernard Cohen and Anne Whitman (London, 1999).

Isaac Newton, *Opticks: or A Treatise of the Reflexions, Refractions, Inflexions and Colours of Light* (London, 1704).

Isaac Newton, *A Short Chronicle from the First Memory of Things in Europe to the Conquest of Persia by Alexander the Great* (London, 1728).

*The Correspondence of Isaac Newton*, edited by Herbert W. Turnbull, John F. Scott, A. Rupert Hall and Laura Tilling (Cambridge, 1959–1977).

*The Mathematical Papers of Isaac Newton*, edited by D.T. Whiteside (Cambridge, 1967–1981).


*Secondary Works*

Gale E. Christianson, *In the Presence of the Creator: Isaac Newton and His Times* (New York, 1984).

I. Bernard Cohen and George E. Smith (editors), *The Cambridge Companion to Newton* (Cambridge, 2002).

Patricia Fara, *Newton: The Making of a Genius* (London, 2003).

John Fauvel, Raymond Flood, Michael Shortland and

Robin Wilson (editors), *Let Newton Be!* (Oxford, 1988).

James Gleick, *Isaac Newton* (London, 2003).

A. Rupert Hall, *Isaac Newton: Adventurer in Thought* (Oxford, 1992).

Alexander Koyre, *Newtonian Studies* (Chicago, 1965).

Frank E. Manuel, *A Portrait of Isaac Newton* (Cambridge, Mass., 1968).

Richard S. Westfall, *Never at Rest: A Biography of Isaac Newton* (Cambridge, 1980).

Michael White, *Isaac Newton, The Last Sorcerer* (London, 1997).

# Index

# BY PETER ACKROYD
## ALSO AVAILABLE FROM VINTAGE

| | | | |
|---|---|---|---|
| ☐ | Blake | 9780749391768 | £9.99 |
| ☐ | Dan Leno and the Limehouse Golem | 9780749396596 | £7.99 |
| ☐ | Dickens (Abridged) | 9780099437093 | £9.99 |
| ☐ | The Life of Thomas More | 9780749386405 | £8.99 |
| ☐ | London: The Biography | 9780099422587 | £14.99 |
| ☐ | Milton in America | 9780749386252 | £6.99 |
| ☐ | The Collection | 9780099428947 | £12.99 |
| ☐ | The Plato Papers | 9780099289951 | £6.99 |
| ☐ | Lambs of London | 9780099472094 | £6.99 |
| ☐ | Chaucer | 9780099287483 | £7.99 |
| ☐ | Albion | 9780099438076 | £17.99 |
| ☐ | Clerkenwell Tales | 9780749386306 | £6.99 |
| ☐ | Turner | 9780099287285 | £6.99 |
| ☐ | Shakespeare | 9780749386559 | £9.99 |

FREE POST AND PACKING
Overseas customers allow £2.00 per paperback

BY PHONE: 01624 677237

BY POST: Random House Books
C/o Bookpost, PO Box 29, Douglas
Isle of Man, IM99 1BQ

BY FAX: 01624 670923

BY EMAIL: bookshop@enterprise.net

Cheques (payable to Bookpost) and credit cards accepted

Prices and availability subject to change without notice.
Allow 28 days for delivery.
When placing your order, please mention if you do not wish to receive
any additional information.

www.vintage-books.co.uk